The Archaeological Survey Manual

The Archaeological Survey Manual

Gregory G. White
Thomas F. King

 Routledge
Taylor & Francis Group

LONDON AND NEW YORK

First published 2007 by Left Coast Press, Inc.

Published 2016 by Routledge
2 Park Square, Milton Park, Abingdon, Oxon OX14 4RN
711 Third Avenue, New York, NY 10017, USA

Routledge is an imprint of the Taylor & Francis Group, an informa business

The archaeological survey manual / Gregory G. White, Thomas F. King,
p. cm.
Includes bibliographical references and index.
ISBN-13: 978-1-59874-008-0 (hardcover : alk. paper)
ISBN-10: 1-59874-008-3 (hardcover : alk. paper)
ISBN-13: 978-1-59874-009-7 (pbk. : alk. paper)
ISBN-10: 1-59874-009-1 (pbk. : alk. paper)
1. Archaeological surveyingóHandbooks, manuals, etc.
2. ArchaeologyóField workóHandbooks, manuals, etc.
I. White, Gregory G. II. King, Thomas F.
CC76.3.A725 2007
930.10286dc22
2006035670

ISBN 13: 978-1-59874-009-7 (pbk)

Table of Contents

Introduction

Equipment

Field Work

Professionalism

Appendices

Introduction

Chapter

Introduction and Definitions

Chapter 1

The dawn is cool and calm as you crawl out of your tent, boots laced, pack in hand. "Coffee," you think, "breakfast, make lunch, clean-up camp, talk to crew," and after three weeks, it's a familiar checklist. At 7:00 a.m. you strike out on a brisk two-mile climb to resume yesterday's survey transect. Today, you'll cover the upslope end of an alpine meadow where it tapers to a narrow notch, the gathering point of an old trail leading to the mountain pass above. Your progress will be slow but fascinating as you identify and record Gold Rush trailside trash scatters, Archaic hunting blinds, and late prehistoric kill sites...archaeological survey!

What Is Archaeological Survey?

Archaeological survey is an anthropological exercise. Anthropology is the study of humankind, and archaeology is the subdiscipline of anthropology concerned with past material culture and the material ramifications of culture. Archaeological survey is a field application consisting of a constellation of methods leading to the detection, identification, and documentation of the material traces of past human activity.

Archaeological survey has grown to incorporate a variety of applications and technical dimensions. With the modern development of remote sensing technologies, archaeological survey is often enhanced by instrumentation and imagery. Some studies result in very little actual outdoor work, with field visits reduced to a targeted examination of particular anomalies. Remote sensing technologies are a productive and growing addition to archaeological survey, but the applications are highly specialized and each may require a separate course of study and career track. In fact, these specialties are increasingly partitioned in real-world practice, with many private, government, and academic professionals rarely leaving the computer lab. There are places in the world where it's not possible to muster a field effort, where budget limitations, logistical obstacles, and even legal, political, or social constraints mean that we cannot get out on the ground. In these cases remote sensing is the end-all for archaeological survey.

However, these cases are rare and the specialties exclusive. The larger picture reveals that conventional field survey is not only the dominant form of archaeological survey, but actually on the increase.

This book is strictly concerned with conventional, on-the-ground, hill-climbing, fence-jumping, rattlesnake-dodging, rock-hopping, tree-ducking archaeological field survey. This book considers issues surrounding the design and implementation of conventional survey and provides training in the recording skills and equipment typically used on conventional survey. We also offer lots of "veteran" tips to quickly bring you up to speed.

Archaeologists are always busy because human beings have always been busy. They left a lot of stuff lying around, and they made and broke things wherever they went. Stuff piled up, was ignored, lost, and tossed, eaten and passed, burned, broken, and buried. People took things from here and put them over there, made mountains and hauled them away, assembled, fabricated, and congealed, and disarticulated, unraveled, and separated practically everything in reach. Traces of past human activity can be found in every corner of the world, and most habitable landscapes are transformed by human dominion.

Once we find them, archaeological traces can provide a mind-boggling array of information about past human behavior. Archaeologists look for these traces, examine their spatial relationships, variations, and functions, date them, associate them with particular cultures, events, or periods, and study how

> ## Here are some things that differentiate archaeological survey from excavation:
>
> » Archaeological survey is logically prior to and complements excavation.
>
> » Archaeological survey is external and horizontal while excavation is internal and vertical.
>
> » Archaeological survey is labor-extensive while excavation is labor-intensive.
>
> » Archaeological survey can produce a distribution pattern reflecting archaeological visibility and past land use practices, while excavation produces a detailed examination of a single landform or living space.
>
> » Survey costs tend to emphasize logistics (travel and housing), while excavation costs tend to emphasize labor (field and especially analysis).

they changed over time. Some archaeological traces are tiny, some monumental, some surprising and unexpected, some prosaic and everyday, but taken as a whole they form the primary record of human activity of all kinds and through all ages. Conventional field survey is the primary means archaeologists use to detect, define, and record these phenomena.

Archaeological Survey Skills and Professional Archaeology

In the United States as in much of the modern world, the practice of professional archaeology is embedded in a complex and rigorous body of national and regional laws and policies that have been enacted to make sure that—prior to development—government and private entities consider the potential effects of their actions on the archaeological record. Within this area of professional practice, known in this country as part of *cultural resource management* (CRM), archaeological surveys are regularly done in tandem with planning and environmental studies in advance of construction and land use projects. Consequently, field survey has grown in importance as a means of identifying and inventorying archaeological resources.

The United States is now well into its second generation of archaeologists who graduated from college and entered a professional CRM environment where most career opportunities are found in private and government practice, and where the most varied and profitable opportunities for research and field work are found outside the academic context. Long dead is the traditional model of archaeological education:

> which may be characterized, only somewhat facetiously, as the training of archaeology students by professors of archaeology to produce more professors of archaeology to train more students (Miller 2000: 69).

This is because there are more roles, wonderful careers, and exciting work all over the map. How much have things changed? In the most recent Society for American Archaeology (SAA) membership survey (SAA Needs Assessment Survey 2003), of 634 survey respondents, approximately 33 percent were academic faculty or staff, while the remaining nonstudent, non-retired respondents were in government or private practice (28 percent) or careers affiliated with academic cultural resource management programs or museum institutions, which straddle the traditional divide (12 percent). State-level archaeological societies show an even stronger trend, with the extreme case represented in California, where a 2000 membership survey (Wheeler 2000, 2001) found membership (averaging around 900) primarily employed in private, professional archaeological practice (48.5%), a nearly equal number employed in federal, state, county, or city professional positions (42.3%), a few describing avocational or tribal affiliation (6.2%), and only a handful in academic employment, which included faculty, staff, or academic CRM positions (3.0%). If California is a bellwether for national trends, then things have clearly changed a lot and will continue to change in the future: the lion's share of jobs for archaeologists are and will be in government and private practice.

This is important because it affects decisions you can reach right now about training and preparation for a career as a professional archaeologist. The most transferable of these skills are the ones most broadly applicable: the basic methodologies, methods, and techniques of field archaeology.

Who Should Read This Book?

This volume is addressed to three major groups of readers. Our primary target reader is the student and avocational practitioner, whose needs and interests are identified above.

Our second target is the professional practitioner. To this group we will be saying nothing new; rather, we will be discussing things that everybody knows but for some reason seldom writes about. By writing about them, as simply as possible in a single volume, we hope to encourage our colleagues to think clearly and critically about the methods they employ, the ways in which they decide which methods to employ in different situations, and the ways they communicate with non-archaeologists about what they're doing and why they're doing it. We also hope to prevail upon archaeologists to think about their practice from the standpoint of those who use their products for management purposes and those who pay the bill for the conduct of surveys.

The third type of reader is the non-archaeological manager—for example, in the U.S., an official in a federal agency, a State Historic Preservation Officer (SHPO), someone who manages land, or someone who plans, finances, or undertakes land development. These individuals have occasion to wonder when it's useful to conduct archaeological surveys, what such a survey ought to entail, and what it may cost. We hope this book will help answer some of these questions, demystify things a bit, and help readers ask better questions of the archaeologists with whom they deal.

About This Book

This manual provides the basic conceptual tools, key methods, and techniques of modern field survey. Archaeology is a practical field that demands practical knowledge and quick decisions about things and places. This manual will help train you to locate,

Here are some things that define and distinguish archaeological survey:

» Archaeological field survey can be used to build a regional inventory of archaeological phenomena.

» Archaeological survey can contribute to the preservation and enhancement of knowledge of previously known resources.

» Archaeological survey can reaffirm our expectations for human settlement and land use by finding sites, features, and artifacts in predictable patterns.

» Archaeological survey can identify unanticipated resources in unexpected places and lead to important new findings on the age, extent, and variability of past human activity.

identify, and record archaeological phenomena, but it should be used in combination with field experience. We are convinced that archaeology should not be taught or learned from books alone. Whether an airy manual or ponderous "shelf-lunker," when it comes to archaeological training a book is a pale companion to experience.

This book is about doing archaeological survey, not "cultural resource survey," and we refer to "archaeological sites" as the phenomena sought, rather than to "historic properties" or "cultural resources." In fact, it is important for students and professionals alike to keep archaeological survey in context as a *part* of "cultural resource" survey. "Cultural Resources" include much more than archaeological sites: they can be old buildings, neighborhoods, landscapes, traditional cultural sites and areas, sacred sites, and even songs, stories, dance forms, and groups of people. It is important to consider all the impacts of proposed projects on all kinds of cultural resources, but finding archaeological sites involves particular methods, which comprise the subject of this book. No one should think, however, that by doing an archaeological survey one has located—or even responsibly looked for—all kinds of cultural resources.

The core of this book began in 1978 with the junior author's tome *The Archeological Survey: Methods and Uses (TASMU)* distributed without update or revision for more than twenty years by the Heritage Conservation and Recreation Service, and its successor, the National Park Service, U.S. Department of the Interior (King 1978), and still available today on-line from the National Park Service On-Line Book service at http://www.cr.nps.gov/history/online_books/king/index.htm. *TASMU* was a popular and widely distributed document, one of the first field survey manuals primarily designed to meet the needs of non-archaeological managers in government and private practice who had little direct understanding of archaeology, yet found themselves pressed by circumstances to find answers to their questions about archaeological survey.

The second component of this book originated in the senior author's *Archaeological Survey Handbook (ASH)* which was distributed at field schools and field classes he taught at California State University, Chico (CSU-Chico). ASH was also used as a reference by staff associated with CSU-Chico's cultural resource management program office. *ASH* was designed to convey the skills necessary for the practice of professional archaeological survey and site recording and was meant to be used by students seeking a professional career. Each manual told half the story. *TASMU* was text-rich and deliberative while *ASH* was graphical and practical.

Thus, in this new manual we seek to meld these themes and address the needs of student, professional, and avocational practitioners looking to sharpen skills and master common practice.

Chapter-by-Chapter

Chapter 2, A Short History of Archaeological Survey in North America, provides a brief overview of the history of archaeological survey in North America, touching on the broad theoretical and legal trends that affected how survey has been conducted, who

surveyed, where, and for what purpose. Through the first half of the 20th century, archaeological survey was done primarily by academicians looking for sites to dig, usually in places where they expected to find big sites that might produce cultural sequences. In the last half of the 20th century, archaeological survey was increasingly done by people in state, federal, academic, and private practice for the purpose of fulfilling the requirements of new historic preservation laws. The sites they found and decisions they faced challenged old definitions and assumptions. Now, in the 21st century, most archaeologists should expect and plan for a career as a cultural resource professional in some nonacademic setting, where archaeological sites may be a small fraction of the cultural resources designated for identification, study, and management.

Chapter 3, Gearing Up for Archaeological Survey, is the first of six chapters that provide detailed instruction on the selection and use of archaeological survey gear. Chapter 3 sets the stage by listing equipment needed on an everyday basis as well as that used in various special circumstances. We also offer veteran tips to help you buy, handle, and maintain your own equipment in comfort and safety.

Chapter 4, Compass Features and Use, offers a series of instructional modules on compass use, including features to look for when you are selecting a compass for purchase, setting and using declination, using and recording azimuth and quadrant notation, sighting a compass bearing, and using a clinometer. The chapter closes with a few of those veteran tips.

Chapter 5, Map Use 1: Basics and Orienteering, covers map basics and offers a series of map reading exercises. We zero in on the types of maps in most common use by field professionals to plan, document, and report archaeological surveys. The chapter describes what these maps represent, their scale, orientation, and grid positions, their basic notations and symbols, and their use in assessing natural and cultural features. The chapter shows how to read topographic lines, use a map and compass to triangulate a position, and use a map and compass to traverse to another location.

Chapter 6, Map Use 2: the PLSS, describes the Public Land Survey System (PLSS), a system established by the Continental Congress in 1785 and used to measure and assess lands annexed to the original 13 states. It is important to recognize and understand PLSS notation because it was used historically to establish homestead, allotment, civic, county, and state boundaries, and is still used by most state and federal agencies as a standard real property description.

Chapter 7, Map Use 3: Coordinate Systems, introduces the two coordinate systems in most common use on archaeological field survey, the latitude and longitude and Universal Transverse Mercator/Universal Polar Stereographic (UTM/UPS) systems. Latitude and longitude are the world's oldest and most widespread comprehensive coordinate system, and like the PLSS, your reasons to learn and use the system involve both historical and modern documentation. The UTM/UPS system came into common use in the late 20th century and is now the principal coordinate system employed by modern computer-based information management.

Chapter 8, GPS in the Field, provides a brief overview of the selection and use of hand-held Global Positioning System (GPS) receivers on archaeological surveys. Recommendations are offered and the most common uses and pitfalls of the technology are described.

Chapter 9, Prefield Research and Survey Design, is the first of five chapters that turn attention to archaeological survey design, using a hypothetical example, an archaeological survey study in Griffin Valley, Indeterminate State. This chapter describes the human past and the history of archaeological investigations in Griffin Valley, then illustrates how different approaches to survey can result in different portraits of the past. It emphasizes the importance of prefield research, biases and bias-control, scope-of-work, sources, prefield agreements, and interacting with stakeholders.

Chapter 10, Types of Archaeological Field Survey, covers the use and abuse of field survey methods. Two basic classes of pedestrian survey are identified, *Exclusive* and *Non-exclusive,* and different adaptations and variations of each are described, emphasizing the connection that must be established between goals and purpose, and the particular sampling biases introduced by each method. Special types of survey requiring unique methodological approaches are identified, and predictive survey methods and sampling schemes are discussed.

Chapter 11, Archaeological Field Survey Methods, turns attention to on-the-ground field methods, including those that pertain to individual crew members and those that apply to team organization and activity. With respect to individual skills, we offer exercises in training the eye to key into surface archaeological phenomena, establishing pace rates, working on a transect team, and methods for assaying possible subsurface deposits. As for organization and tracking, we provide tips on transect methods, crew management, field notes and records, and survey tracking and quality control. The inventory forms and log forms presented in *Appendix C: Example Archaeological Survey Tracking and Organization Forms*, and *Appendix D: Example Archaeological Survey Field Notes* are designed to support Chapter 11, and provide models of forms in common use by professional archaeologists.

Chapter 12, Adapting Methods to Purpose, returns to our hypothetical example and focuses on how our surveyors use prefield research to help them decide how to design an archaeological survey in Griffin Valley, and then how they implement the archaeological survey in specific environmental settings. The chapter ends with a warning against prescriptive methods and an exhortation to think things through.

Chapter 13, Archaeological Site Documentation, offers tips on what to do once you have encountered an archaeological site. We divide the tasks into three areas that are usually sequential—assay, documentation, and quality control—with findings from each determining actions taken at the next stage. Team members engaged in assay determine site boundaries, establish site datums, locate artifacts, and assay the subsurface extent of the site. On the basis of these findings, team members can then document, including creating a site sketch map, filling out a site record, securing site imagery, and completing additional records, logs, and forms, as appropriate. The chapter closes with comments on the importance of making quality control a goal while you are still in the field. The site record form offered in *Appendix B: Example Archaeological Site Record Form* is designed to support Chapter 13, and is a distillation of forms prescribed and recommended by state and federal agencies throughout North America.

Chapter 14, Being Responsible, is a brief treatment of a complicated subject, in which we identify the common pitfalls encountered by professionals—rookie and veteran. We discuss our obligation to report in a timely and thorough manner, and recommend that you recognize and act within the limits of your own training and the scope of archaeology within the field of cultural resource management. We also offer tips on how to stay current. The URLs offered in *Appendix A: State Archaeological Survey Standards, Guidelines, and Forms Websites* are designed to support Chapter 14. These are direct links to—as far as we can tell—the most current survey standards available for each state (including some .doc and .pdf files). Some of these standards are promulgated as recommendations, some as state law, some advanced by offices of historic preservation (or the equivalent agency), and some by the state archaeologist. Some of these links will also lead you directly to credentialing, licensing, or other state-level investigator standards.

The chapter, and book, closes with a discussion of our basic obligations to be respectful and communicate with stakeholders, especially descendant communities. We advocate including these communities at all stages of the field effort, in prefield research, field identification and assessment, and in crafting analyses and recommendations after the field work is complete.

A Short History of Archaeological Survey in North America

Chapter 2

Modern archaeology in North America has roots in the dilettantism and antiquarianism of the 19th century. Some of the earliest "archaeologists" were explorers, traveling journalists, soldiers, and natural scientists who described archaeological sites that they sought out or stumbled upon during travels in the little-known regions of the West and Latin America (i.e., Stephens 1841). In this sense, archaeological survey has a long tradition in American archaeology.

Some of the earliest archaeological publications were essentially survey volumes, describing the ruins or mounds that had been discovered in some particular area of the continent, discussing collections that had been derived from them, and speculating on their origins and functions (i.e., Squier and Davis 1848). These studies were a far cry from the systematic surveys conducted by archaeologists today; they were simple explorations in which the fieldworker described those phenomena that came to his attention with no pretense of identifying all the vestiges of past human activity in the area. Such full descriptions were not necessary to the authors' purposes.

As archaeology became a recognized discipline in the United States in the latter part of the 19th century, such general exploratory surveys were a normal part of its research repertoire. The purposes of survey were almost totally descriptive. Sites might generally be compared and contrasted with one another on the basis of survey data, but in most cases the survey was regarded primarily as a prelude to excavation. One surveyed to locate sites to dig. Survey methods were not the subject of much concern. The archaeologist presumably knew what kinds of sites he wanted to dig, and the survey simply involved looking for them. If other sorts of sites were missed in the process, this was of no importance, because the archaeologist did not want to dig them anyway. The survey was not itself seen as a research tool, since relationships among sites were not generally considered important.

At this time, archaeology was primarily oriented toward the study of change in artifact types, structural types, and other attributes of archaeological sites through time. The aims of such studies were the characterization of modern or recent groups in time-depth, the search for the origins of particular cultures, and the reconstruction of culture history (Willey and Sabloff 1974:42–64). Early studies were directed toward demonstrating what were thought to be universal patterns of human cultural evolution, showing that given societies had advanced inevitably through stages equated with "savagery" and "barbarism" to "civilization."

In the early 20th century, the concept of unilinear cultural evolution began to go out of vogue, to be replaced with what became known as "historical particularism." Historical particularism denied the possibility of readily demonstrating large-scale evolutionary changes. Particularists argued instead for the painstaking reconstruction of the histories of particular peoples and cultures. These small histories, it was thought, could eventually be synthesized to permit the development of an understanding of cultural evolution in general. Archaeologists trained in the historical particularist tradition naturally tended to direct their research toward the reconstruction of the culture history of a particular site or small area. Very careful study of local culture-change sequences became the rule of the day (Willey and Sabloff 1974:88–98).

The physical focus of study for the particularist archaeologist, however, was not greatly different from that for the unilinear evolutionist. Both approaches caused archaeologists to value large, deep sites with many strata, or other indicators of change through time, and with many artifacts that could be equated with things used by groups that still functioned at the time, or that had been documented as more or less functioning entities by historians, cultural anthropologists, sociologists, and travelers. In such sites, cultures could be speculatively described through analogy with living groups that used similar artifacts and could be seen succeeding one another in more or less orderly progression through the successive soil strata. When surveying an area, most archaeologists sought sites of this type and virtually or entirely ignored the small, shallow, or recent sites that did not promise to contribute directly to culture history reconstruction.

During the 1930s, American archaeology became deeply involved in the emergency employment programs initiated by the Roosevelt administration. Large numbers of workers employed by the Works Progress (later Projects) Administration (WPA) or Civilian Conservation Corps (CCC) could be committed to archaeological activities, to do socially useful work under relatively low-cost supervision. As a result, huge crews were thrown together and sent into the field under the leadership of archaeologists— themselves often young graduate students or avocational archaeologists (Heizer 1974; Lyon 1982). While some of these projects were utter disasters, others provided extremely important bodies of data, and the exercise had profound effects on the nature of archaeological practice. One such effect was on archaeological survey.

Large areas were surveyed, in advance of construction projects such as reservoirs or simply in order to deploy large numbers of people in activities that would do minimum damage to archaeological sites. Because the workers were unskilled and their supervisors often not highly trained or broadly experienced, it was necessary to develop standardized methods of recording sites. Work was usually undertaken in areas where large populations of unemployed workers existed, not necessarily where an archaeologist, left to his own devices, might have chosen to work. In consequence, the archaeologists often found themselves dealing with areas they did not know well, where they were not sure just what kinds of sites to seek. Thus it became necessary to think about what constituted an archaeological site, what made such a site important, and to record a wider range of sites than would have been recorded by an archaeologist simply seeking sites to dig for pure research purposes.

The rationale for survey remained the discovery of sites for excavation, and reconstruction of culture history was still the main reason for excavation. Changes were in the making, however, springing in part from other effects of the Depression and its archaeological activity. The make-work programs had often forced archaeologists into work they would not have done otherwise, at sites that would normally not have tempted them. Large work crews made possible the stripping of large site-areas, revealing the organization of entire prehistoric villages and showing that more could be studied about extinct human groups than the ways their artifacts changed through time. The Roosevelt administration, with its socialist policy overtones, may have helped set the stage— together with the worldwide academic excitement about Marxism—for the rise of cultural materialism in anthropological theory, which characterized the 1960s. This in turn would contribute to a major change in the ways in which archaeologists looked at their world and their research base.

Before these changes took place, however, archaeology lapsed into general quiescence during World War II. After the war, with the initiation of huge water-control projects across the nation, archaeology was faced with a major challenge, and the era of "salvage archaeology" began in earnest. Initially, the Smithsonian Institution and the National Park Service undertook the river-basin salvage program; after passage of the Reservoir Salvage Act of 1960 (Public Law 86-523), the responsibility became more and more concentrated within the National Park Service until, in 1969, the Smithsonian Institution divorced itself from the program entirely.

Surveys were obviously required as the first step in most reservoir salvage projects. These surveys were still aimed almost exclusively at locating sites to dig, and sites were still chosen for excavation primarily when it was thought that they would contribute to the construction of cultural-historical sequences. The scant funds appropriated to the National Park Service for salvage were largely reserved for excavation, so surveys were done cheaply and fast. The results were sometimes appalling by modern standards. At the proposed New Melones Reservoir in California, for example, a river-basin salvage program survey in 1948 recorded only four sites, none of which was regarded as being of sufficient importance to justify expenditure of the program's limited salvage money (Fredrickson 1949). Based on surveys in the 1970s, the reservoir area (where construction had not yet gotten underway) was recognized as a National Register District because of its 190 prehistoric and over 400 historic sites, structures, buildings, and objects (Moratto 1976).

When the interstate highway system went into construction during the 1950s, archaeological surveyors faced a new set of challenges. Lacking the congressional mandate for a salvage program like the one provided for reservoirs, archaeologists developed such arrangements as they could with their state highway agencies for the conduct of surveys and salvage. Pipeline construction companies also began to make arrangements for salvage. One of the first and most successful highway and pipeline salvage programs was that developed by Fred Wendorf and his colleagues at the Museum of New Mexico. Wendorf's experiences resulted in the preparation of a manual on salvage archaeology. The conditions under which many salvage surveys were done were described as follows:

> The archaeological teams follow as closely behind the surveyors and as far ahead of the right-of-way clearing machinery as possible. Even under ideal conditions the timing will still be close, and there may not be more than three to four weeks between the survey and dozer clearing the right-of-way. (Wendorf 1962: 54)

Working under such pressures, archaeologists found themselves having to think in new ways about old and little-considered questions. What sites were worth recording? What made a site worth digging? How could they locate, record, and excavate them most efficiently?

As the 1950s progressed, anthropologists in many of the nation's universities had become dissatisfied with historical particularism as their basic approach to understanding human society (see Harris 1968 and Garbarino 1977 for extended discussions). Generally, the construction of local culture histories had not provided the basis for syntheses that revealed much about culture change. Some began to believe that the cart had been placed before the horse, i.e., that precise questions should be developed about culture change and hypothetical answers proposed, before data could be collected in a fruitful manner. It was recognized that it was physically impossible to collect all the data that might exist about any living culture or any archaeological site; data collection is always selective. Without having formulated questions to be answered

before field work began, it was unlikely that the anthropologist or archaeologist in the field would select the data necessary to answer them. At the same time, with the decline of the extreme anti-communism of the mid-50s, theories that purported to explain culture change using models derived from Marx and other materialist thinkers became popular. Testing materialist propositions about the nature and causes of culture change required the study of relationships between human society and the material resource base—the natural environment. The rise of environmental anthropology (Steward 1955; White 1959) coincided nicely with the growing recognition among salvage archaeologists that there were scientifically valuable data in sites that were not large or deeply stratified. In fact, it was clear that if one really wanted to understand the relationships between human groups and their environments, one needed to look at all kinds of sites, representing all kinds of interactions with the environment. Small sites representing a small range of activities carried out during a single season with reference to a single economic resource were at least as important to understanding human-environment relations as were big, deep sites where people lived repeatedly or year round and engaged in a diversity of activities.

Not only did small, shallow sites begin to attract attention from environmentally oriented archaeologists; the relationships among sites, and between constellations of sites and the environment in which they existed, became popular subjects for study. The focus of archaeology during the 1960s shifted rapidly, from the individual site to the regional settlement pattern (i.e., Chang 1968). Archaeological survey itself began to be recognized as an important research tool in its own right, one which, even without associated excavation, could show how human populations and their activities had been distributed within a natural environment. Survey was gradually redefined, no longer viewed simply as exploration to find sites to dig but as a systematic effort to "provide information on the number, the location, and the nature of the sites within a given region" (Heizer and Graham 1967: 14).

In 1966, congress enacted the Reservoir Salvage Act which directed the National Park Service to oversee archaeological survey and salvage excavation within the flood zone for all new and planned reservoirs funded, constructed, or licensed by the federal government. While the Reservoir Salvage Act also extended the Historic Sites Act of 1935, survey and salvage requirements were not named for other types of federal undertaking. Nevertheless, the Reservoir Salvage Act had an immediate and major impact on the known archaeological record of many mountainous regions. The remote and often wilderness lands generally planned for water development had been subject to little or no formal archaeological survey work in the years before 1966. With implementation of the Reservoir Salvage Act, a myriad of planned or potential water projects produced a wealth of new, comprehensive survey data (e.g., Childress and Chartkoff 1966; King 1966; Larrabee and Kardas 1966; Moratto 1972; Turpin 1991). By design these projects focused on river canyons and alluvial basins, and this meant that upland sites were not inspected. However, new kinds of sites were brought to the attention of archaeologists and managers all over the U.S. Decisions about which sites would be dug, recorded, removed, or sealed thrust many "new" and "old" archaeologists into the doomsday triage of salvage archaeology, highlighting the issue of relative significance by forcing the scientists to make critical decisions about labor investment under salvage conditions.

The importance of archaeological survey as a research activity continued to grow during the 1960s and 70s, but the literature concerning survey methods remained surprisingly limited. In 1966, Reynold Ruppe published a case study concerned with demonstrating how well-organized, problem-oriented archaeological survey could "be made to produce information that is usually considered procurable only by excavation" (Ruppe 1966: 331). Many studies based on survey data were published during the decade between Ruppe's work and the first publication of this volume (i.e., Matson and Lipe 1975; Schiffer and House 1975, 1977; Thomas 1975), all devoted at least in part to the advantages and deficiencies of particular survey strategies, or to comparing strategies (i.e., Lovis 1976; Mueller 1974, 1975). In subsequent years, articles dealing with survey methodology appeared regularly in the professional literature (i.e., Aldenderfer and Schieppati 1984; Aldenderfer and Pierce 1984; Gallant 1986; Redman 1982; Schiffer et al. 1978), most discussing specific methods such as shovel testing and the application of geophysical methods, or on data management via geographic information systems. A few general archaeological methods texts have included sections or even whole volumes on survey (i.e., Ammerman 1981; Collins and Molyneaux 2003; Hester et al. 1997).

The great change that has taken place since publication of the junior author's *The Archeological Survey: Methods and Uses* in its first, 1978 iteration has been a proliferation of surveys conducted in response to CRM laws and regulations. It has come to be pretty widely accepted that archaeological surveys are needed during planning for projects that may disturb the ground, and large numbers of more or less trained archaeologists are employed in conducting such surveys, usually under contract to change agents like government agencies and private sector development interests.

The Promulgation of Archaeological Survey Standards

In many parts of the United States and in other countries, concern has developed about the quality of survey work performed under contract. This has typically resulted in the promulgation of survey standards of the kind identified in Appendix A. Such standards are usually developed by a government agency (e.g. a state historic preservation officer) or by a group of practicing archaeologists (e.g., a state or regional archaeological council).

Promulgation of such standards is an understandable expression of the very real need for comparability when judging survey results. When a reviewer with some kind of management responsibility receives two survey reports, he or she naturally wants to know that both authors meant something roughly comparable by the word "survey." It is impossible to judge the accuracy of survey results if "survey" to one person means driving past a parcel and sweeping it with binoculars, while the same word to another means crawling over it on hands and knees. Budgeting for survey and judging the adequacy of budgets attached to survey proposals are impossible in the absence of some agreement about what "survey" means.

But the downside of standards is that they standardize, limiting the flexibility one needs in order to deal with varying circumstances, problems, and opportunities. And they tend to restrict creativity, imposing modes of behavior that are acceptable to the majority, or simply to those with the power to impose the standards. They are often adopted—or so it seems to us—without much thought about their purposes and implications, as simple reflections of "the way we've always done it around here" or of "the way we think it ought to be done in the best of all possible worlds."

When applied to real-world situations, such standards do not always work very well; they can generate unnecessary costs, frustrate creativity and cost-effectiveness, and result in the systematic failure to note certain kinds of phenomena or to address particular needs.

For example, most state and regional survey standards devote a good deal of attention to how far apart archaeological surveyors should walk when inspecting the ground and to whether and how far apart they should dig shovel-test holes. These standards may produce nothing but frustration when applied, say, to a valley that has experienced heavy recent siltation, to a range of rocky hills, or to a suburban neighborhood. At the same time, most standards say little or nothing about seeking out and interviewing artifact collectors and others with knowledge of archaeological site locations. Such people may have spent decades walking and observing the land under different and changing circumstances and may be able to say far more about the area's archaeological sites than can be discovered via short-term inspection of the ground.

Strict standards are also most easily applied to tightly defined areas, such as the actual footprint of a proposed construction project. It may be virtually impossible to apply them to softer, more loosely defined areas, like the areas where the indirect impacts of a project may occur. If one is devoted to the application of hard and fast standards, one may be tempted to ignore situations to which such standards can't be easily applied—focusing, for example, only on areas of anticipated direct, physical, immediate impact and ignoring indirect effects altogether. Yet the indirect effects may be the truly important effects of a proposed project, which should not be ignored in planning.

A fundamental assumption of this book is that there is no standard way to do archaeological survey. Different methods and approaches are appropriate for different circumstances and to achieve different purposes. The challenge in planning a survey is to decide what one really needs to identify and the amount of detail one needs to record about what one identifies, and then to design the most efficient, effective means of identifying it.

Because it is often forgotten, by both archaeologists and those who employ them, we end with another reminder that doing a good archaeological survey may more or less guarantee that one will identify all the archaeological sites in the area one is concerned about (for example, the area to be affected by a development project), but it will not guarantee the identification of all cultural resources. Other kinds of expert studies and consultation with a variety of stakeholders are necessary to identify such cultural resource types as historic buildings, tribal sacred sites, traditional landscapes, culturally important plants and animals, and valued aspects of community life. Where archaeological survey is necessary as part of cultural resource identification, however, it should be carefully and thoughtfully planned and carried out, using whatever methods are appropriate to the kinds of sites and situations expected in the area under study.

We'll return in Chapter 10 to how a range of survey standards can be employed to address different situations. First, however, we'll turn to the basic equipment and skills you need to master before you head into the field.

Equipment

Chapter

Gearing Up for Archaeological Survey

Chapter 3

It's time to gear up. This and the following four chapters describe the necessary equipment and all-important orienteering skills you'll need to conduct a professional archaeological field survey. Use the equipment checklist on page 19 to stock your personal gear or use it as a packing-list reminder before you head into the field.

Most or all the equipment listed below is available on-line at competitive prices. A web search for "archaeology equipment" will probably not produce very satisfying results. You'll get more hits on the web if you search for "equipment for environmental professionals," and there are a couple very good catalog suppliers of outdoor engineering equipment.

Don't automatically buy based on price, and don't assume expensive is the best. Check with archaeological survey veterans for preferred brands and models. Most of archaeology's peculiar equipment traditions are built out of solid experience with brands and models that have earned reputations for reliability.

Here are some equipment tips. Please see chapters 4 through 8 for tips and recommendations on map compasses, GPS Units, and USGS topographic maps.

Footwear

Shoes are not listed in the checklist on page 19, but they are supremely important; your feet are your most important piece of equipment. Wear boots, and select and prepare them wisely. We recommend classic, high-top, lace-up boots that cover the ankles. A couple reasons: (1) they support the ankles and (2) rattlesnakes. Break them in long before you start field work. Carry an extra pair of socks, extra boot strings, and "moleskin" (a thick but soft cotton or flannel

adhesive padding that helps prevent or relieve blister pain). Your first week out, cover the whole back of your heel with moleskin. In other words, make sure you put the moleskin on before you get a blister, not after. This stuff has a firm adhesive so it can survive a couple showers or swims. Cut the patch with rounded corners and it will not catch and peel off when you pull on your socks.

Pack or Vest

Should you buy a vest or stick with your old pack? The pack and vest are roughly equivalent, and sometimes the choice is a matter of economics or logistics: economics because some packs are less expensive than vests, which are pretty high-end (more than $60.00); logistics because packs carry a larger load a bit easier on a long trek to a backcountry survey area.

We like packs for bulky or heavy items such as screens, folding shovels, collapsible bucket augers, datum stakes, and signage. Take stock of your survey needs and assign one or more people to wear packs, if necessary. Otherwise, we like vests because they do what they're designed to do, keep constant-use items right up front within easy reach.

Packs are not as useful for everyday hand-held gear because you have to take off the pack and unzip a pocket. It's not unusual to see the datum at a newly encountered site adorned with three or four disgorged packs and the owners tramping back for tools every time the tasks change. A vest usually solves this and keeps its owner mobile and efficient.

One drawback common to nearly all commercial vests is the infuriatingly maladaptive design of the *Big Back Pocket* (BBP). BBPs might be used to store bulky items like stakes and screens, but they are not really designed for this purpose. These things should just go in a pack. The real problem is everyday storage. The BBP is where you want to store your clipboard, lunch, and water, and the problem is that most BBPs don't do this very well. The best vests have side-slot access to the BBP so it's not necessary to remove the vest and all you have to do is reach back to pull out your clipboard. Unfortunately, most packs with this useful feature also have side slots that are too thin to accommodate port-a-desk-type clipboards, which we recommend here.

Clipboard and Binder

The aluminum portable desk-style clipboard is in common use in the environmental sciences and has a critical function on archaeological survey (see Chapter 13). We prefer a side-opening style with a fold-out writing plate and snap-in form holder in the well. If the well is deep enough, you can also hold recording tools such as pencils, erasers, scales, a ruler, and templates. However, buy for comfort—the deeper the well the harder it is to hold the clipboard comfortably. You should be able to find a useful clipboard at your local engineering supply store. However, shop around for the best price: we recently searched on-line for "aluminum clipboard" and kicked up hundreds of excellent alternatives in the $20.00 to $40.00 range.

We recommend keeping a small (1-inch spine) three-ring binder in your vest or pack loaded with empty sheet protectors to store, protect, and organize your finished draft field records and to hold your notebook, extra graph paper, maps, and identification guides.

Writing Instruments

Use a pencil, not a pen, to complete forms, sketch maps, and notes. Most professionals prefer mechanical pencils over traditional, wood-encased graphite pencils because the lead can be controlled more readily, and the barrel of the pencil can be adjusted for comfort and stay unchanged throughout the life of the instrument (doesn't get sharpened away).

Mechanical pencils have one big bugaboo, and that's the eraser. A dysfunctional eraser stub can produce hours of frustration and the senior author has been known to go totally *Cave Man* when one of those little eraser dealies busts off. Plus, the cap that covers gets lost immediately, so you end up gumming-up the

eraser with hand contact and snapping it off when you click the lead. Spare yourself this experience and look around before you select a pencil. The newest mechanical pencils solve all these problems by using a long, refillable eraser insert, and advancing the lead with one motion (end click) and the eraser with another (shaft twist).

Pencil lead (really, processed graphite and clay) is important when it comes to record-keeping of all sorts. Leads range from soft to hard, thin to thick. There are notation systems for both hardness and thickness, and you'll need to know these systems before you buy. Taking hardness first, in the standard American grade school system pencil hardness ranged from soft (#1) to hard (#4). However, most mechanical pencil retailers use a modified European system, with hardness ranging from H (hard) to B (black, i.e., soft) and HB (in-between) then ranked numerically, with moderately to very hard (2H to 9H) and moderate to very soft (2B to 9B).

Generally, the softer 3B to 9B leads are not good for archaeological record-keeping. The softer the lead, the more residue it leaves on the form or sketch map, and this can be a problem because work done in soft lead is easily smudged and streaked. However, 3H to 9H leads are also problematic. While it is true that the harder lead is more resistent to smudging, you have to press too hard into the paper to get a strong, visible line, making it easy to snap a fine lead or poke and rip the paper. The solution is an HB lead (equivalent to the good old #2), which makes a bold line and minimizes smudging.

As for lead sizes, we recommend keeping two pencils, one with 5 mm lead and one with 9 mm lead, because these options will give you a lot more latitude drawing distinctive symbols and features on site sketch maps (see Chapter 13).

It's a good idea to carry a soft engineering gum or plastic eraser. You'll also find it handy to carry a permanent ink felt-tip pen to mark flagging, pin flags, and field collection bags.

Recording Tools

Make sure you carry a metric/English scale ruler, tape measure (5 m/16 ft), and reel tape (30m/100ft). You are likely to use both metric and English scales. It is common practice in most states to record historical features in English rule and prehistoric features in metric (standard) rule.

A protractor is an essential tool for triangulation (Chapter 5) and sketch map production (Chapter 13). If you carry just one protractor in your toolkit, make it a simple full ring (360°). When you select and buy your protractor, be conscious of different kinds of bearing notations and the need to match your compass and protractor formats. Generally, a standard full-ring protractor will have azimuth degree markers, with an inner set running east and an outer ring running west. This is what you will generally need and use—an azimuth protractor matching an azimuth compass (see Chapter 4 for azimuth notation details).

If you plan to work in a tropical zone—or during the rainy season in any environment—you should acquire a supply of rain-proof paper and writing instruments, such as Rite in the Rain notebooks, papers, and pens. The notebooks resist wrinkling and ink running, and the looseleaf paper can be loaded into a photocopy machine or printer to produce custom rain-proof recording forms.

A north arrow/scale template appears on page 20. Use this when you take photos of features and site datums (Chapter 13). Two additional templates are essential to the well-rounded archaeological field survey toolkit: the Land Area Template (Chapter 6) and the UTM Grid Template (Chapter 7). Recommendations for graph paper appear in Chapter 13. Appendices to this book include an example site recording form (Appendix B), example survey tracking forms (Appendix C), and example field notes (Appendix D).

Hand Tools

Trowels are not exclusive to excavation, and have regular and important function on survey, too, such as excavation of shovel test pits, surface scrapes, and clearing around a datum. Trowel selection is steeped in many mysteries and grand archaeological traditions. However, when it comes to survey, make sure you have a trowel you won't mind losing. Trowels are forgotten and left behind, but more often, when you survey in woody or brushy terrain, a branch whipping back behind you can grab a trowel by the crook in the handle and throw it a long, long way. They can be propelled over cliffs, shot into impenetrable brush, or

tossed silently and undetected into oblivion. On page 20, we offer an old trick to help you keep that trowel.

A small, solar-powered calculator is very useful on field survey for calculating pace and coverage rates, and for estimating site area. A timepiece is also an essential safety and organizational tool.

Field communications are a problem in large coverage areas or with a dispersed crew. In this situation you should strongly consider assigning each crew member a standard playground sports whistle or a two-way radio. If a whistle, you'll need to agree on signals (e.g., "log-in" = three short blasts). If two-way radios, you'll find that powerful and simple "walkie-talkies" are easily procured. Make sure you agree on a channel and make sure you check out local zoning restrictions, security considerations (near military bases, especially), and restrictions on use of local channels around radio-controlled explosives.

A hand lens or pocket magnifier can be very useful for identification of artifacts or mineral attributes. It is common to see veteran surveyors attach a hand lens to their compass lanyard.

The crew chief should carry and be responsible for the use of a safety kit. In addition to the standard issue for such a kit, you should add large bandages (including "butterfly" bandages) and a roll of moleskin.

To round out your field equipment, you will also find daily and obvious uses for a pocketknife, toilet paper, a flashlight, canteen, and lunch container/cold pack. An extra pair of socks is often the difference between a miserable and beautiful day.

The Package

Arrange your gear for comfort and efficiency. A full pack is not so much a sign of a well-prepared archaeologist, as an invitation to muscle strain. You won't need all the stuff listed here on every survey, and of the stuff you really need, most of it can be distributed among crew depending on their roles and responsibilities. Think it through and pick from this list.

Most field archaeologists build a toolkit over the span of a career, and many veterans carry gear that is polished by time and hard use. Experienced archaeologists purchase gear only after long deliberation and tend to its maintenance with great care. It is not hard to find an archaeologist who possesses a truly goofy fetish for a favorite trowel or compass. Don't even think about asking to borrow someone's favorite gear. You should be willing to take time to find the perfect pack or vest. It will take time to fill it with the right combination of useful items.

Gear Up!: Assemble Your Essential Survey Equipment

✪ Constant Use

- ☐ Day Pack or Survey Vest
- ☐ Map Compass
- ☐ GPS Unit
- ☐ USGS Topographic Maps
- ☐ Three-Ring Binder and Sheet Protectors
- ☐ Timepiece
- ☐ Whistle
- ☐ Two-Way Radio
- ☐ Trowel
- ☐ Canteen
- ☐ Pocketknife
- ☐ Toilet Paper (OK, not constant)
- ☐ Flashlight
- ☐ Safety Kit
- ☐ Lunch Container
- ☐ Cold Pack

✪ Tests and Probes

- ☐ Small, Hand-Held Screen
- ☐ Folding Shovel
- ☐ Collapsible Bucket Auger
- ☐ Soil Color Chart
- ☐ Grain Size/Roundness Gauge

✪ When You Find a Site

- ☐ Clipboard
- ☐ UTM Grid Template
- ☐ Land-Area Template
- ☐ Graph Paper
- ☐ Recording Forms
- ☐ Field Journal/Notebook
- ☐ Mechanical Pencils
- ☐ Ink/Pencil Erasers
- ☐ Metric/English Scale Ruler
- ☐ Protractor (full ring)
- ☐ North Arrow/Scale
- ☐ Calculator
- ☐ Hand Lens/Pocket Magnifier
- ☐ Metric/English Locking Tape Measure
- ☐ Metric/English Reel Tape (100 ft/30 m)
- ☐ 35mm or Digital Camera
- ☐ Pin Flags

✪ Delineation

- ☐ Surveyor's Flagging
- ☐ Datum Stakes
- ☐ Antiquities Warnings
- ☐ Witness Signs

Equipment Tips

For Your Own Comfort and Safety. When weather, bugs, and vegetation permit, wear a light, longsleeve cotton shirt, a comfortable straw hat, sun block, and lip balm.

Buy a high high quality 5" or 6" square or pointed trowel.

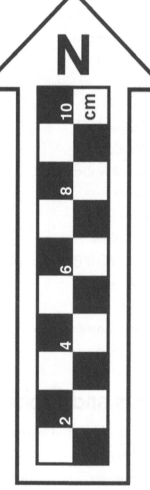

When you are hiking, slip the blade under your belt and the tip into your rear pocket. The latter will prevent the trowel from flipping out of your belt if the handle is caught by a branch.

N

10 cm

8

6

4

2

Scan and print-out this North Arrow/Scale template on card stock. Use it when you photograph feature details.

Compass Features and Use

Chapter 4

This and the next three chapters focus on basic orienteering skills using graphical presentations and exercises, beginning here with compass features and use. The compass is used in all phases of archaeological survey, and few tools are as important and basic, with key roles in plotting location, aligning transect coverage, recording survey observations, and making site sketch maps.

The first step is to select a compass, and here the advice is easy: archaeological survey needs are best served using a "map compass" (p. 22). Make sure you select a compass with these minimum features:

(1) a revolving compass housing (p. 22).

(2) a 0–360° azimuth ring (p. 22).

(3) declination adjustment (p. 23).

(4) azimuth east or both azimuth east and west bearing notation (p. 24).

(5) prismatic sighting (p. 26).

(6) a clinometer (p. 27).

All these features are necessary to take standard bearings and angles as detailed in the following pages.

You should also look for other useful features, and stay away from some not so useful. For example, most map compasses have a hinged sighting lid, but some of these lids secure with a snap catch and some do not. Get one with the snap catch because it does a better job keeping the compass closed to the elements.

Try to find a compass with a see-through bed because it will be easier to use for triangulation or other orienteering functions involving maps.

However, some compasses are made with a shaded green or yellow translucent bed, and these are problematic because they act like lenses, filtering out map symbols.

Some compasses are made with embossed USGS map scales and map orientation lines and these can make it easier to use the compass for map measurement functions that would otherwise require a separate template. However, it is rare to find a compass with a useful scale—one that matches the scale on the map. Some of the scales come close but don't quite match (e.g., many compass map scales are 1:25,000 while the map scale is 1:24,000).

Some compasses have built-in magnification, but this is not especially useful because these magnifiers are poor quality and tend to get scratched up with regular use.

Compass Exercises

Once you select a compass, turn to the following pages for exercises in seven basic skills:

(1) identify compass features;

(2) set compass declination;

(3) review azimuth notation;

(4) review quadrant notation;

(5) sight a compass bearing;

(6) take a clinometer reading; and,

(7) triangulate your position using a compass and a USGS quad sheet.

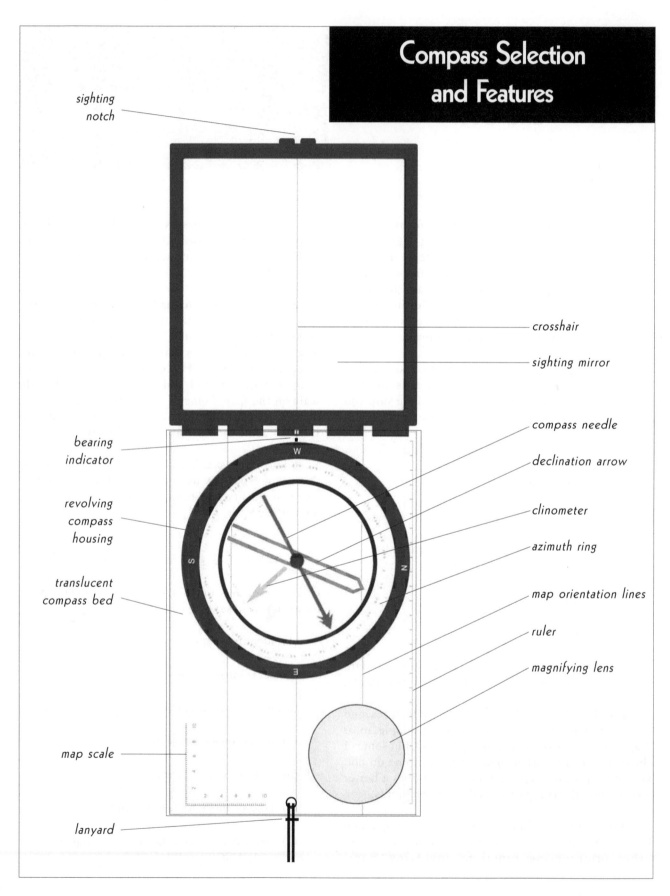

Compass Selection and Features

sighting notch

crosshair

sighting mirror

compass needle

bearing indicator

declination arrow

revolving compass housing

clinometer

azimuth ring

translucent compass bed

map orientation lines

ruler

magnifying lens

map scale

lanyard

The compass needle is magnetized and when it swings freely it will point to magnetic north. However, move the compass near an object of iron or steel, and the needle will respond.

Compass Declination Adjustment

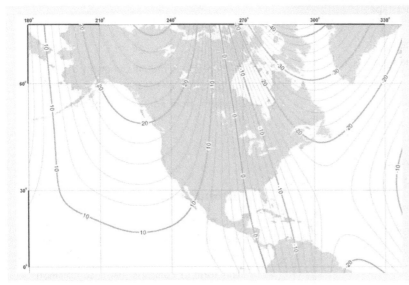

Similarly, declination is a manifestation at the surface of large-scale magnetic variation affected by forces associated with the earth's magnetic field, mass, and mineral composition. In the Midwest, the compass needle will point between 0° to 10.0° east or west of "true" (geographic) north.

adapted from NOAA (http:/ /www.ngdc.noaa.gov/seg/ WMM/data/wmm-D05.pdf)

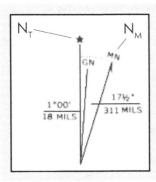

Land area maps and measurement systems are keyed to true north, requiring a compass adjustment. Magnetic declination in your study area is indicated at the bottom of the USGS quad sheet. This example plots true north (N_T) N17½°E of magnetic north (N_M).

To adjust declination, use the tab-shaped screwdriver attached to the lanyard.

West of the 0° line, declination correction is easterly, requiring a clockwise adjustment of the declination arrow, i.e., N_T is east of N_M.

Adjust the declination arrow until it lines up on the correct easterly bearing.

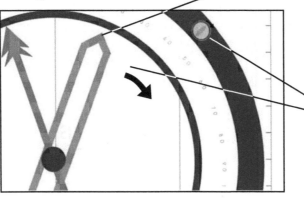

Turn the screw clockwise to move the declination arrow easterly, opposite for westerly.

Types of Compass Dials: Azimuth Notation

Before you purchase a compass, take a close look at the dial. There are many different types of dials, and each produces a distinct type of reading. The three most common dial types are azimuth east, azimuth west, and quadrant.

Azimuth Notation is the measurement of a horizontal position, bearing, or angle along a full, 360° perimeter from a fixed zero point.

degrees increase to the east

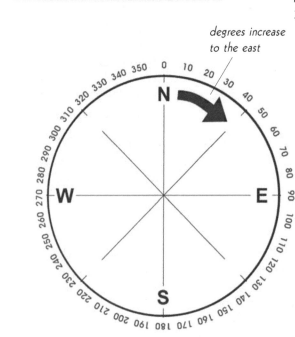

Azimuth East

The compass face in most common use is the azimuth east, where North is 0° and east 90°, south 180°, and west 270°. This form of compass dial is so common that many fieldworkers assume all compasses are like this and automatically write down the bearing as a simple number (e.g., "168°") without looking at or making note of the specific dial type.

degrees increase to the west

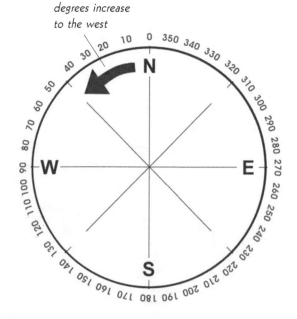

Azimuth West

The azimuth west compass is less common but often sold off the shelf and is hard to differentiate without close inspection. The dial is divided into 360°, North is 0° and south 180°, but numbers increase west of north so that west is 90° and east 270°.

Quadrant notation divides the compass face into quarters.

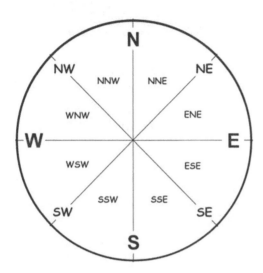

Types of Compass Dials: Quadrant Notation

Directional Compass

A directional compass is used to find an approximate horizontal position relative to the four cardinal points. *Quadral bearings* include the four cardinal points (N, S, E, W) and the four mid-points (NW, NE, SE, SW) represented by northeast, southeast, southwest, and northwest quadrants. *Octal bearings* specify position relative to two sub-coordinates per quadrant (e.g., NNW, NNE, ENE, etc.).

Quadratic Compass

A quadratic compass establishes polar zero points and measures horizontal position, bearing, or angle east and west of zero north or zero south. For example, due northeast on a directional compass is N45°E on a quadratic compass. Due east may be N90°E or S90°E and due west may be N90°W or S90°W, depending on your preferred default pole.

Quadratic compasses were common in the past. Historical engineering records and the journals of early mining and exploration are often based on quadratic compass notation.

Quadratic compass notation is still preferred by some who like its capacity for easy traversing and transpositioning. For example, N45°E is exactly opposite S45°W, and so on.

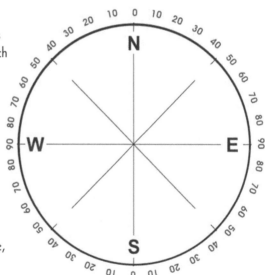

Use the Full Expression

Because these is so much variation out there in compass dial types, never write down a bearing as a simple number only (e.g., 127°). Always use the full expression. A bearing notation should be trinomial (three parts): datum, angle, direction (e.g., N127°E). Watch the following pages for examples.

Azimuth East Notation	Azimuth West Notation	Quadrant Notation
≤360°	≤360°	Strictly ≤90°
N___°E	N___°W	N___°E (or) N___°W
		S___°W (or) S___°E

Sight a Compass Bearing

Sighting a bearing is the most basic, fundamental, and common operation you will encounter using a compass. Practice until you are confident in this skill.

First: Aim
Aim the compass at an object or topographic feature.

Second: Align
Align the sighting notch with the target, and align the crosshair through the center of the reflected image of the compass dial.

Fourth: Read
The bearing is indicated at the top of the dial, in this case N18°E.

Third: Turn
Now turn the revolving compass housing until the declination arrow boxes-in the compass needle.

Use the Clinometer

The clinometer measures <u>degree</u> of slope, not the percent or angle of slope.

Turn the compass dial until either "W" (N270°E) or "E" (N90°E)is aligned with the bearing indicator.

The clinometer needle hangs free and points to the degree of slope.

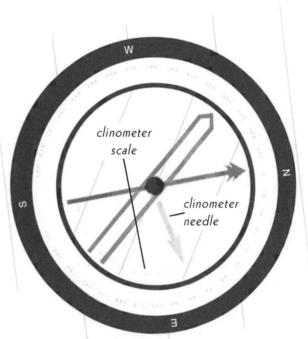

clinometer scale

clinometer needle

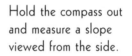

Hold the compass out and measure a slope viewed from the side.

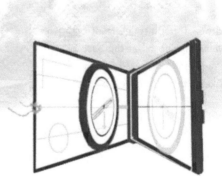

Or, aim the compass uphill and shoot in an object at about your height above the ground. Use the sighting mirror to read the clinometer.

Compass Tips

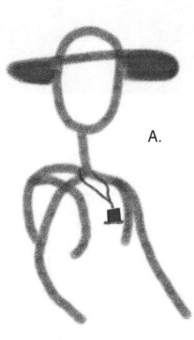

A.

B.

When you're in the field, it's important to have the compass out and ready for use all the time. You can loop your compass lanyard around your neck (A), but this makes it hard to hold it out far enough in front of your face to take a bearing. To solve this problem, hitch your compass lanyard to the shoulder of your vest or pack strap using a slip knot or hasp (B). Some vests even have a grommet on the shoulder strap for just this purpose. In either case, keep the compass in your top vest or shirt pocket to keep it from banging around as you walk.

A drop of graphite lubricant or sewing machine oil at the base of the revolving compass housing will keep the action free.

When you aim and align remember to bend the lid, don't crane you neck!

Map Use 1: Basics and Orienteering

Chapter 5

Maps are fundamental to all phases of investigation, from project planning to spatial interpretation of findings. Maps are especially important in the field for tracking, orienteering, and documentation. The most commonly used maps are published by the United States Geological Survey (USGS) and are called "quad sheets"—short for quadrangle or four-sided maps. USGS has accomplished complete coverage of the U.S. in two detailed series: 1:62,500 and 1:24,000 scale maps. The scales represent the relationship between a measured map distance and a physical or real distance on the landscape. In the 1:62,500 maps, one inch on the map represents 62,500 inches real distance (about 1.0 mile). This, is a fairly coarse-grained map, and archaeologists prefer finer detail on natural and constructed features so the 1:24,000 (1 inch=0.38 mile) map is favored wherever it is available.

The 1:62,500 maps are also known as "15-minute" (15') quads because they frame an area measuring 15' of latitude and 15' of longitude. In turn, the 1:24,000 maps are often identified as "7.5 minute" (7.5') quads because they frame an area measuring 7.5' latitude and 7.5' longitude (p. 53). The 7.5' quads display a quarter of the space contained in a 15' quad but in a larger format, at a much smaller scale, and with considerably more detail on localized landscape variation. Together these quad sheets form a continuous mosaic, matching side-to-side, end-to-end (p. 30) with amazing detail and accuracy.

The contiguous U.S. states and Hawaii are generally covered by one or more series of older, often historical 15' maps, in some cases 50 to 130 years old, and these have been replaced in the last 10 to 40 years by the finer-grained 7.5' quads. For Alaska, there are some 7.5' maps around urban areas such as Fairbanks and Anchorage, otherwise, only the 15' series has been published.

Most of the map-use exercises presented below are pared down enough to make a geographer turn that special shade of crimson reserved for their encounters with the unremediable foolish few. We acknowledge the shortcomings but point out that this is an introduction to map use for field archaeologists who will use maps in the peculiar circumstance of day-to-day conventional archaeological survey. Our exercises focus on basic navigation techniques such as traversing and triangulating, land measurement functions, and the tools and techniques used to identify and document points, alignments, and areas.

There are, however, lots of resources out there for those who want to pursue the geometric and geospatial basis of map-making. For example, we recommend that the reader follow up on these exercises by visiting map instruction websites hosted by USGS. The agency publishes on-line tutorials and a number of "Fact Sheets" that cover basic map conventions and provide additional useful details. You can find an interesting introductory topographic map tutorial on-line at the USGS website (http://erg.usgs.gov/isb/pubs/booklets/topo/topo.html). On-line Fact Sheet 190-95 provides a list of map indices available from USGS (http://data.geocomm.com/sdts/fs19095.pdf). You can also search for and order USGS maps on-line (http://topomaps.usgs.gov/ordering_maps.html). A useful tutorial on topographic map scales is also available at the USGS website (http://mac.usgs.gov/isb/pubs/factsheets/fs01502.html).

We provide topographic map symbol charts on pages 39–42, in black-and-white. Because it is easier to discriminate some of the symbols by reference to a full-color chart, we recommend you also consult the USGS topographic map symbol charts at:

http://erg.usgs.gov/isb/pubs/booklets/symbols/

Map Grid

Each quadrangle sheet is 7.5' latitude and 7.5' longitude. Together they form a continuous mosaic, matching side-to-side, end-to-end, with amazing detail on natural and constructed features.

The three quads east of this sheet are:
- Tuscan Buttes NE
- Dales
- Tuscan Springs

Identifying Conjoining Maps

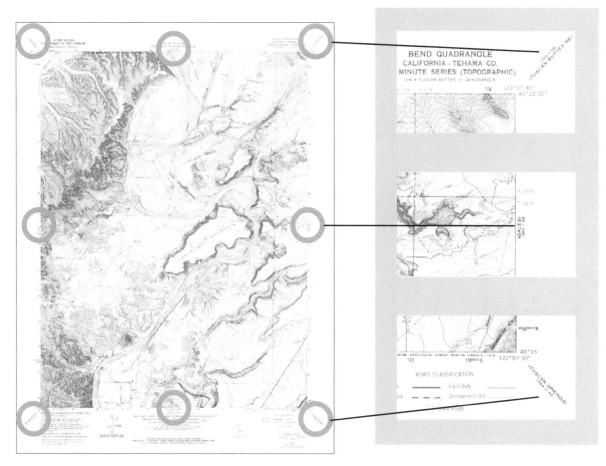

On recent sheets, quadrangle location is indicated with a grid.

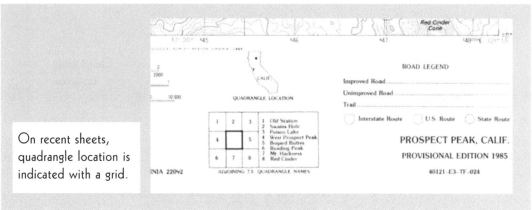

Map Information

Map conventions, publication information, map key, and coordinate information are published at the corners and along the bottom margin.

Map datum and declination tend to be the most important, day-to-day.

Upper Left

UNITED STATES
DEPARTMENT OF THE INTERIOR
GEOLOGICAL SURVEY

122°15'
30"   164000m E   65

Map Publisher

USGS 7.5'
Quadrangle

Upper Right

BEND QUADRANGLE
CALIFORNIA–TEHAMA CO.
7.5 MINUTE SERIES (TOPOGRAPHIC)
SW/4 TUSCAN BUTTES 15' QUADRANGLE

0 000 FEET R X W   R 6 W   574   122°07'30"
                                                40°2

Map Name and Type

Publication, Datum, and Revision Dates

2°15'   14 MI. TO INTERSTATE 5   65
          84 MI. TO RED BLUFF

Mapped, edited, and published by the Geological Survey

Control by USGS and USC&GS

Topography by photogrammetric methods from aerial
photographs taken 1963   Field checked 1965

Polyconic projection.   1927 North American datum
10,000 foot grid based on California coordinate system, zone 1
1000-meter Universal Transverse Mercator grid ticks,
zone 10, shown in blue

Fine red dashed lines indicate selected fence lines

Certain land lines are omitted because of insufficient data

Revision shown in purple compiled from aerial photographs
taken 1976 and other source data.   This information not
field checked.   Map edited 1978

Road Symbols, Map Name, and Series Identification

INTERIOR—GEOLOGICAL SURVEY, RESTON, VIRGINIA—1978
                                                    40°1
573   574000m E   122°07'30"

ROAD CLASSIFICATION

Heavy duty ━━━━━━ Light duty ━━━━━━
Medium duty ━━━━━━ Unimproved dirt ┄┄┄┄┄┄

○ State Route

BEND, CALIF.
SW/4 TUSCAN BUTTES 15' QUADRANGLE
N4015—W12207.5/7.5

1965
PHOTOREVISED 1978
AMS 1564 I SW—SERIES V895

Lower Left   *Lower Right*

Declination Indicator

Map Scales, Contour Interval, and Datum Information

Quadrangle Location

1 940 000 FEET   467   12'30"   468   (RED BLUFF EAST)   478   10'   469
                                              1580 II NW

SCALE 1:24000

CONTOUR INTERVAL 10 FEET
DOTTED LINES CROSSING RIVERS REPRESENT 5 FOOT CONTOURS
NATIONAL GEODETIC VERTICAL DATUM OF 1929

GN
1°76'
311 MILS

UTM GRID AND 1978 MAGNETIC NORTH
DECLINATION AT CENTER OF SHEET

QUADRANGLE LOCATION

THIS MAP COMPLIES WITH NATIONAL MAP ACCURACY STANDARDS
FOR SALE BY U.S. GEOLOGICAL SURVEY, DENVER, COLORADO 80225, OR RESTON, VIRGINIA 22092
A FOLDER DESCRIBING TOPOGRAPHIC MAPS AND SYMBOLS IS AVAILABLE ON REQUEST

Bottom Center

Contours show the land area contained in an elevation interval. Elevations are measured above mean sea level (AMSL).

On USGS topographic maps, contour lines are depicted with brown lines.*

Contour Lines

Contour Spacing:

Tightly spaced contours indicate a steep slope.

Widely spaced contours indicate a shallow slope.

Flat land is represented by few or no contours.

Types of Contour Lines

Index (Primary) Contour
Bold, thick line that plots key intervals (e.g., 100 ft, 50 m). Labeled with elevation measurement above mean sea level.

Intermediate Contour
Light, thin line that plots the regular contour interval (e.g., 40 ft, 10 m)

Supplementary Contour
Added to show existence of slighht topographic features otherwise missed by registered contour intervals.

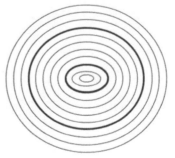

Depression
Added to maps to depict hollows, basins, and pans

This illustration shows a simple conical form depicted in shaded perspective and contour.

* underwater (bathymetric) contours are depicted with blue lines.

Contour Intervals

Contour intervals vary depending on the regional trend in topographic variability, with flatland regions rendered using 10-foot contours and mountainous lands drawn using up to 100-foot or 200-foot contours.

For map makers, the idea is to convey the most information about topography without making the map too "busy." As shown in the Stone Valley example, below right, a tight, 10-foot contour interval in even moderately hilly terrain can create an overdone and almost unreadable map.

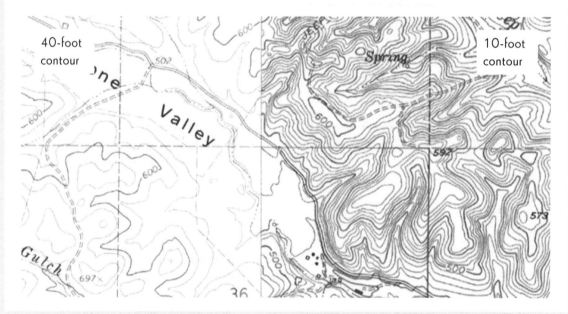

40-foot contour

10-foot contour

Many quad sheets show 40-foot intervals. In this case, every fifth contour (100 feet) is an "index contour."

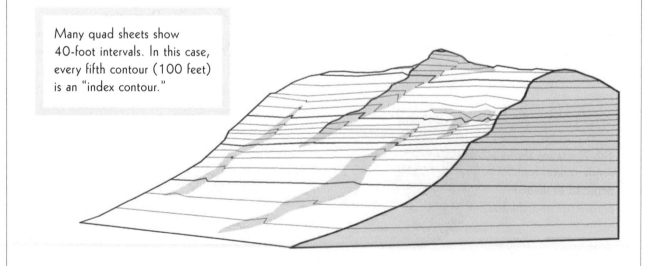

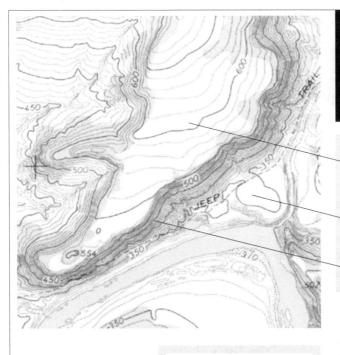

Landform Recognition

A gently sloping table-top mountain bordered by very steep bluffs.

An old, elevated alluvial terrace sculpted by erosion.

A near vertical cliff face.

An alluvial basin topography heavily dissected by active and extinct river channels, including islands, oxbows and sloughs.

New river channel (photorevised).

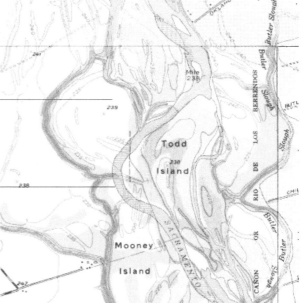

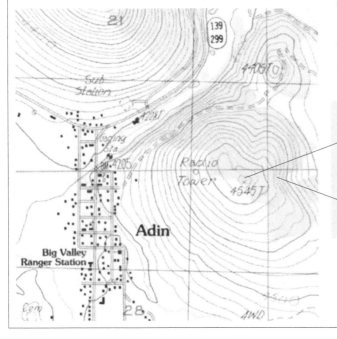

A prominent conical hill overlooking a small town.

The north slope of the hill is forested.

Traverse to a Known Location

To find your way to a location plotted on a topo map:

First: Identify your current location on the topo map based on observation of local features or triangulation (p. 37).

Second: Identify your destination on the topo map, for example, these named stream falls.

Third: Use a ruler to draw a long, straight, true north "zero" line (N0°E) through your current location.

Fourth: Center your protractor over your position, intersecting the zero line at 0° (top/north) and 180° (south/bottom).

Fifth: Using the ruler, draw a line from your position to your destination, and read the bearing where it intersects the protractor, in this case N147°E.

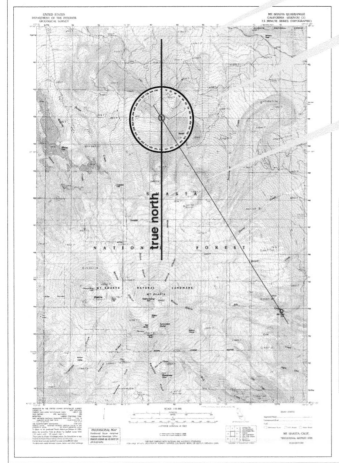

Sixth: Set your compass dial to the bearing, then holding the compass away from you, turn until the north arrow is boxed in by the true north indicator. Now, you should be facing your destination.

Use the map scale to estimate distance. As you proceed, repeat step five to confirm your direction.

To pinpoint your current position on a topo map:

Triangulate Your Location

First: Observe two features on the terrain that can also be identified on the topo map, for example, these two prominent ridge peaks.

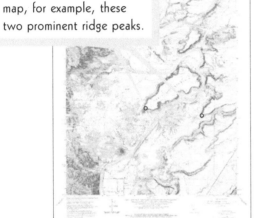

Second: Holding your position, sight bearings to each feature. Make sure your compass is adjusted to local declination (see p. 23).

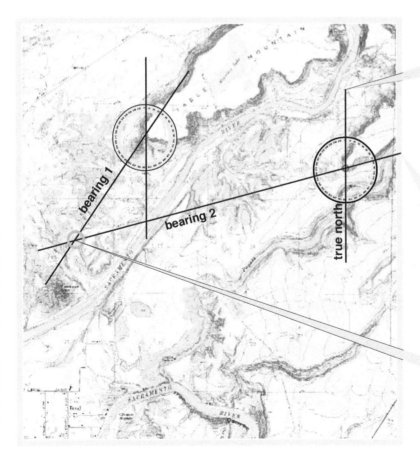

bearing 1

bearing 2

true north

Third: Use a ruler to draw a long, straight, vertical line through each feature representing a true north "zero" line (N0°E). This line will be parallel to the side borders of the map.

Fourth: Use your protractor to plot the two bearings on the topo sheet as straight lines drawn through the features relative to the zero line.

Fifth: The intersection of the two lines pinpoints your position.

Fold Your Topo Map for Field Use

Color printouts from digital map programs are great, but nothing beats a full sheet topo map for print quality and ease of use.

The map is not a piece of art, it's a tracking tool, so feel free to use it like a tool. Pencil in your notes on coverage, site locations, and other observations. Fold the map for ease of use.

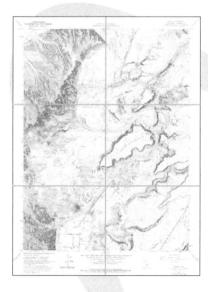

First: Start with a new map, and trim about three-quarters of an inch from top, bottom, and both sides, but leave all map information.

Second: Fold in half lengthwise, map face out.

Fourth: For long-term storage, finish with the map title on top.

Third: Fold in thirds lengthwise; finish with your work area on top.

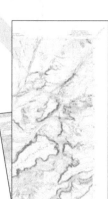

The folded map measures 8.5x10 in.

Map Symbols: Boundaries and Monuments

These symbols depict federal, state, county, or municipal boundary lines, markers, corners, and elevations.

Key: (bk) = black; (rd) = red; (bw)=brown; (bu)=blue; (gr)=green

— — · — National Boundary (bk)

▬▬ · · ▬ State Boundary (bk)

▬ · ▬ · ▬ County Boundary (bk)

▬ · ▬ · ▬ City Boundary (bk)

— — — Town Boundary (bk)

· · · · · · · Small Park Boundary (bk)

▬▬▬ National Park (bk/rd)

— · · — □ Land Grant/Claim (rd)

— · — · — Fence Line (rd)

------- Fence Line (bk)

——— Township Line (rd)

——— Section Line (rd)

WC + MC Witness/Meander Corner (rd)

Neace
△ Third Order or Better Marker (bk)

BM
△ 45.1 Third Order or Better, with Elevation (bk)

△ 19.5 Checked Spot Elevation (bk)

Cactus △ Marker at Section Corner (bk/rd)

+ Unmonumented (bk)

BM × 16.3 Third Order or Better, with Tablet (bk)

× 120.0 Third Order or Better, Recoverable Marker (bk)

BM + 18.6 Bench Mark at Found Section Corner (bk/rd)

× 5.3 Spot Elevation (bk)

BM ⊡ 21.6 Boundary Monument with Tablet (bk)

⊡ 171.3 Boundary Monument without Tablet (bk)

67 ⊡ 301.1 Boundary Monument with Number and Elevation (bk)

▲ U.S. Mineral or Location Monument (bk)

Map Symbols: Constructed Features

These symbols depict constructed point features.

Key: (bk) = black; (rd) = red; (bw)=brown; (bu)=blue; (gr)=green

 Flooded Crop (bu)

 Dam (bk/bu)

 Quarry/Open Pit (bk)

 Borrow Pit (bk)

 Adit (bk)

 Prospect (bk)

 Tailings (bw)

 Tailings Pit (bw)

 Vineyard (gr)

 Orchard (gr)

 Cemetery (bk)

 Fill (bw)

 Buildings (bk)

 Buildings (pr)

 School (bk)

Church (bk)

Campsite/Picnicking Area (bk)

Track (bk)

Airport (bk)

Landing Strip (bk)

 Well/Windmill (bk)

 Tank (bk)

Gauge (bk)

Labeled Landmark (bk)

Map Symbols: Roads, Trails, Lines, and Rails

These symbols depict constructed linear features.

Key: (bk) = black; (rd) = red; (bw)=brown; (bu)=blue; (gr)=green

Standard Gauge and Station (bk)

Standard Gauge, Multiple Track (bk)

Abandoned Track (bk)

Narrow-Gauge Track (bk)

Transmission Line (bk)

Telephone Line (bk)

Aboveground Pipeline (bk)

Underground Pipeline (bk)

Levee (bw)

Flume (bu)

Bridge (rd/bu)

Two-Lane Highway (rd/bk)

Two-Lane w/Median (rd)

Under Construction (rd)

Primary Highway (rd)

Secondary Highway (rd)

Improved Surface Road (rd)

Improved Surface Road (bk)

Unimproved Road (bk)

Unimproved Road (bk)

Trail (bk)

Map Symbols: Boundaries and Monuments

Quad sheets offer good, but dated and highly generalized information on natural habitat. Here are the most common symbols for natural features.

Key: (bk) = black; (rd) = red; (bw)=brown; (bu)=blue; (gr)=green

River (bu)

River Rapids (bu)

Intermittent River (bu)

Perennial Stream (bu)

Disappearing Stream (bu)

Small Falls or Rapids (bu)

Intermittent Stream (bu)

Springs (bu)

Perennial Lake (bu)

Intermittent Lake/Pond (bu)

Dry Lake (bu/bw)

Wash (bw)

Lava (bw)

Scrub (gr)

Forest/Woodland (gr)

Marsh (bu)

Submerged Marsh (gr)

Wooded Marsh (bw)

Submerged, Wooded Marsh (bu/gr)

Subject to Inundation (bu)

Glacier or Permanent Snow Field w/ Contours (bw/bu)

Intertidal Rocks (bu/bk)

Sandy Shoals (bu/bw)

Sand (bw)

Gravel (bw)

Map Use 2: The PLSS

Chapter 6

By order of the Continental Congress in 1785, a comprehensive Public Land Survey System (PLSS) was established for the purpose of management of lands in the public domain for homesteading, lease, and sale. The PLSS was designed to exclude several significant blocks of land: navigable waters, the original thirteen states, some land grants in the southwestern U.S., and some Indian reservations. The two former were covered by a preexisting legal system and the latter had no planned allotments.

In the 19th century, the PLSS was used to demarcate and generate legal descriptions for new western territories acquired by the U.S. government. Acknowledging that many and varied systems of Native American land tenure already existed in these regions, and excepting those lands previously held by other adjudicated governmental powers and agreed upon by treaty, the PLSS was the first centrally administered system used to measure, describe, and divide the new lands. The PLSS conventions generally persist to the present day as the basis for real property descriptions. The PLSS introduced rigid standards for the division and legal definition of lands, creating a grid composed of an orderly arrangement of squares and rectangles. For example, lands opened in the northern Plains by the Homestead Act of 1862, homesteads, roads, town locations, and county boundaries were laid out based on the PLSS, creating checkerboard patterns still evident by air. Thus, a working knowledge of the PLSS may help you identify and interpret historical documentation pertinent to your project area. Equally important, it will help you file accurate descriptions of survey coverage and archaeological locations.

Spain, Mexico, France, and Great Britain all issued prior land grants in U.S. territory. A land grant is an area of land to which title was conferred by a predecessor government and confirmed by the U.S government after the territory in which it is situated was acquired by the United States. These lands were never part of the original public domain and were not subject to subdivision by the PLSS.

USGS quad sheets contain many symbols registering PLSS features (pp. 39–42), including land grant boundaries, township and range boundaries, section lines, section corners, witness and meander corners, and markers of various sorts (generally points recognized by signage or witness monuments), and most importantly, bench marks (generally marked by durable monuments, often brass caps fixed in cairns in a poured concrete base). A few important definitions: For each "goodly portion of land" an *initial point* was established, ideally located at the peak of a prominent natural monument. From the datum established at the initial point, a "star-true" line was determined by astronomical observation of a bearing to Polaris, the North Star. Based on the star-true line, two trunk lines were established, a true north-south (longitudinal) *prime* or *principal meridian* and a true east-west (latitudinal) *Baseline*. Intermediate correction lines were then set, with east-west *standard parallels* at 24-mile intervals parallel with the Baseline, and north-south *guide meridians* at 24-mile intervals parallel to the meridian. Formal land measurement and description was based on a grid of 36-square-mile north-south *township* and east-west *range* lines (pp. 45–46) which in turn framed a grid of 1-square-mile *section* lines (p. 47).

PLSS Home

The PLSS on public lands is currently maintained and regulated by the U.S. Bureau of Land Management (BLM). The BLM's Land Survey Information System (LSIS) is the official government website for the distribution of PLSS information and standards:

http://www.geocommunicator.gov/GeoComm/lsis_home/home/index.html

PLSS Coverage

There are a total of 37 initial points established in the 30 states covered whole or in part by the PLSS.

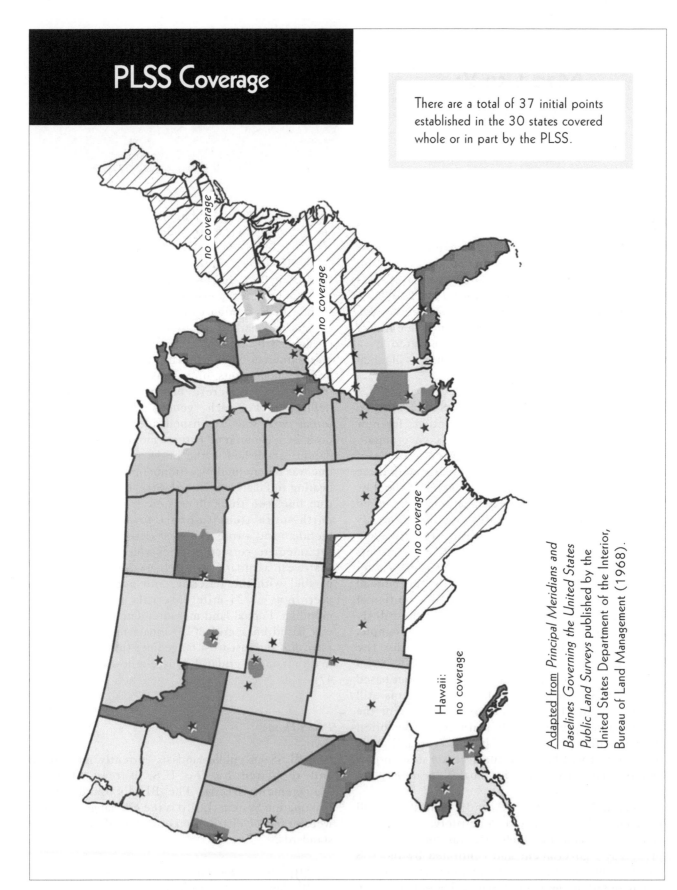

no coverage

no coverage

no coverage

Hawaii:
no coverage

Adapted from *Principal Meridians and Baselines Governing the United States Public Land Surveys* published by the United States Department of the Interior, Bureau of Land Management (1968).

The baseline and prime meridian anchor an additional grid of "correction lines" consisting of guide meridians and standard parallels at 24-mile intervals and township and range lines at 6-mile intervals.

Baseline, Prime Meridian, and Correction Lines

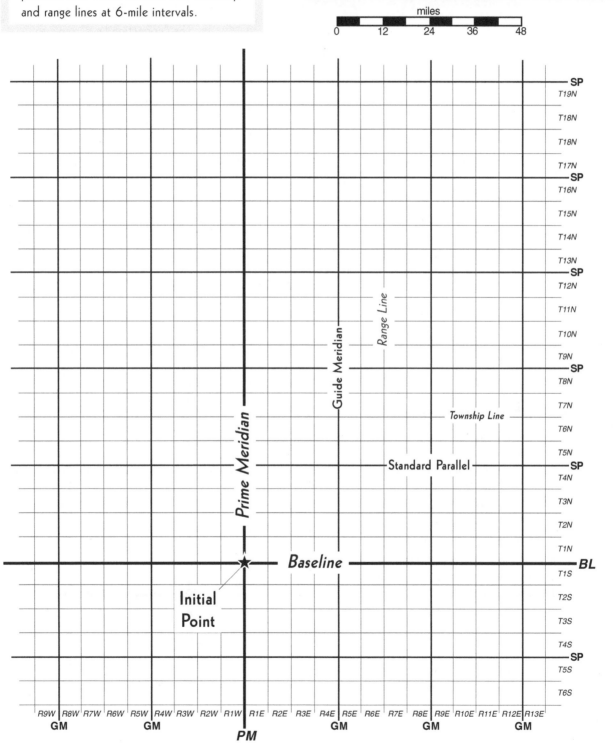

Correction Lines and Township/Range Identification

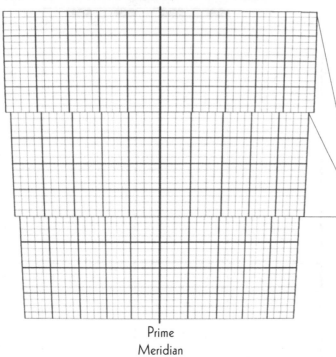

Correction Lines

The "correction lines" serve to correct errors resulting from adapting the grid to the curvature of the earth and to extreme topographic variation associated with mountainous lands.

Correction Lines

Prime
Meridian

Township and Range

Within the 24-mile by 24-mile correction-line blocks, a further subdivision is made consisting of 6-mile-wide east-west township *rows* and 6-mile-wide north-south range *columns*. The townships are numbered consecutively north and south from the baseline, and the ranges are numbered consecutively east and west from the prime meridian. The townships and ranges divide the land into 36-square-mile monumented blocks.

A township (T) and range (R) block is identified based on its grid address, township first. The gray block here would be identified as: T2N/R3E.

Township Line					
6	5	4	3	2	1
7	8	9	10	11	12
18	17	16	15	14	13
19	20	21	22	23	24
30	29	28	27	26	25
31	32	33	34	35	36

Range Line

Sections, Subsections, and Aliquot Parts

Sections

Each Township is further subdivided into 36 one-mile square blocks called "Sections." Each section contains 1 square mile or 640 acres "as near as may be." Sections are numbered consecutively starting with the northeast corner, running east-to-west-to-east-to-west, and so on.

Quarter-Sections

In the physical description of a point or parcel, the section is often further subdivided into 160-acre quarters. Quarter-sections, shown at right, are labeled by their relative quadrant positions. Assuming this is a diagram of the section in gray (above) we would identify the upper left quarter as "the NW¼, Section 22."

NW	NE
SW	SE

NW° of Section 22			
			NE° of NE° of Section 22
EΩ of SW° of Section 22		WΩ of NW° of Section 22	NΩ of NE° of Section 22
		SW° of SE° of Section 22	

Aliquot Parts

An "aliquot part" is any defined, usually square or rectangular, subsection of a section generally a quarter of a quarter, half of a quarter, or a quarter of a quarter of a quarter.

The smallest quarter subdivisions in common use are the quarters of quarters of quarters (10 acres). In descriptive form and on site and survey records these are nested and listed in order from smallest to largest. At left, the parcel shown in gray would be identified as the "NW¼ of the SE¼ of the SE¼ of Section 22."

Find Township and Range on the Quad Sheet

Township is listed on the left and right margins of the map.

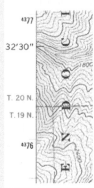

In this example, the mapped area is north of the baseline so townships are north (N) and township numbers increase from bottom to top.

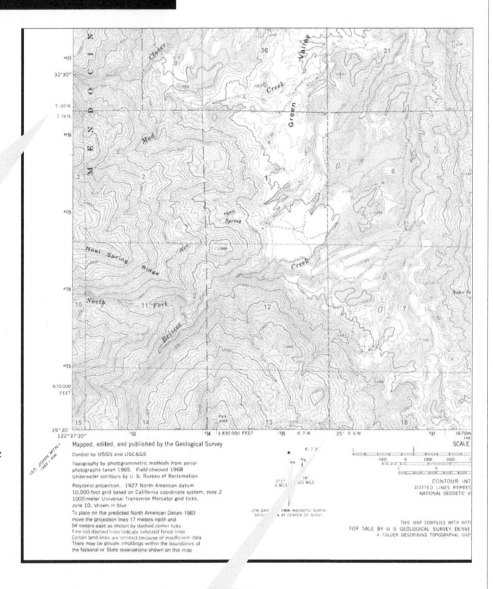

Here, the mapped area is west of the prime meridian, so ranges are west (W) and range numbers increase from right to left.

Range is listed on the top and bottom margins of the map.

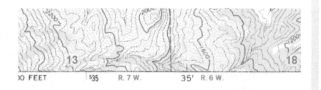

Make Your Own Land-Area Template

Land-area templates are useful tools for determining and describing the location of a parcel or archaeological feature.

You can make your own land-area template. Load a photocopier with transluscent acetate sheets (overlay sheets). Set the copier on 100% and photocopy this page. Cut out the template (below) and use a small die punch to cut the corner holes indicated.

Over the long term you should visit your local engineering supply store and ask for a land-area template. Most of the commercial versions are good, relatively inexpensive, and made from a more robust and durable plastic. However, you should avoid the color-shaded templates because they obscure the map.

Note that the template below is set-up for 7.5' (1:24,000) maps. If you plan to use 15' maps, set the photocopier for 50% reduction. Everything stays the same except the slope indicator scales on the lower right, which should instead read "80-ft contour" and "100-ft contour."

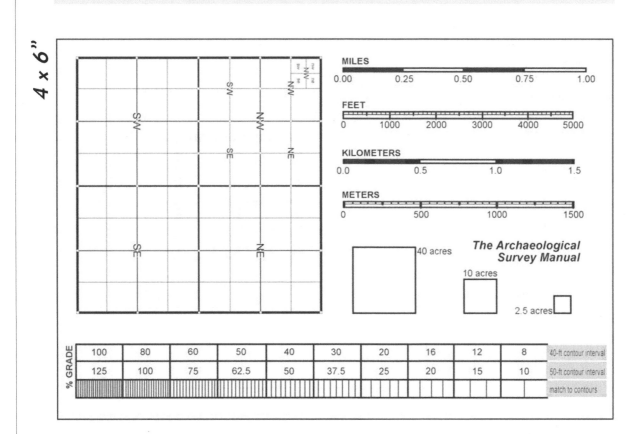

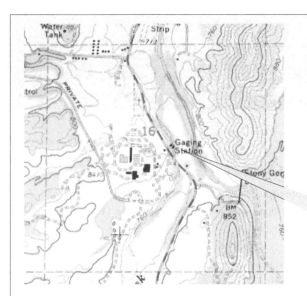

Using the Land-Area Template

First: Identify a feature and determine its township, range, and section address. For example, this stream gauge is located in Township 20N/Range 6W, Section 16.

*Section boundaries adapted to erratic terrain are often odd-shaped polygons. Pencil in an aliquot grid if necessary.

Second: Fit the land area template to the Section boundaries, matching corners as closely as possible.*

You may need to pencil in dots at the subsection corners using the punch holes in the template, but you'll probably master this quickly and make the call without penciling in a grid.

The gauging station is located in the:

SW¼

of the

SW¼

of the

NE¼

of Section 16.

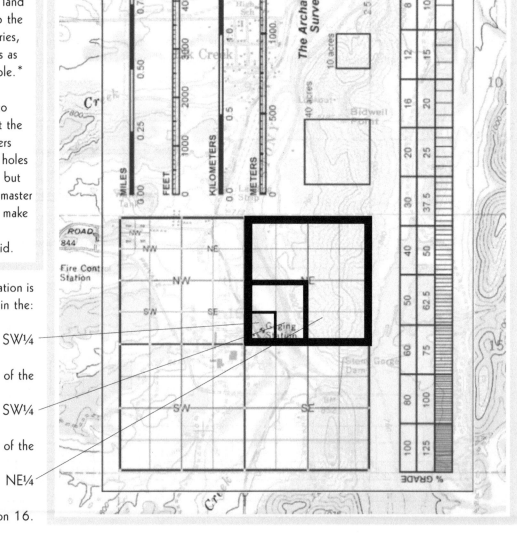

Map Use 3: Coordinate Systems

Chapter 7

Coordinate systems can be used to find, identify, and plot positions, areas, and alignments in horizontal space. Two coordinate systems are in common use on archaeological surveys: latitude and longitude, and Universal Transverse Mercator (UTM)/Universal Polar Stereographic (UPS).

Latitude and Longitude

Latitude and longitude represent the world's oldest and most widespread comprehensive geographical coordinate system. Latitude and longitude coordinates are rarely required in today's archaeological records but there are still several reasons to learn the system. First, the more common PLSS system described above and UTM coordinate system described below are based on a latitude and longitude organizing principle and it is important to know this underlying framework. Second, latitude and longitude are often the only coordinates listed on older site records. Thus, you may need to understand these coordinates in order to find old sites and convert the coordinates to modern standards. Third, latitude and longitude is a common reference system in historical land records and documents such as trail logs, so you will need to know the system in order to locate and track features mentioned in these old sources.

Latitude and longitude positions are described in degrees, minutes, and seconds, each a progressively more precise measure of space. Each degree (°) of latitude or longitude is subdivided into 60 minutes ('). Each minute is subdivided into 60 seconds ("). Specific points, such as site locations, are sometimes measured down to the nearest second but more often to the nearest 20" interval.

In the mid-latitudes of the continental United States, one degree of latitude equals approximately 69 miles (111 km), one minute is just over one mile, and one second is around 100 feet. These distances vary from the equator (where all lines of latitude and longitude diverge to the maximum extent) to the poles (where all lines of longitude and latitude converge).

UTM/UPS

The UTM/UPS system was adopted in 1947 by the U.S. Army and is in common use by recreationalists and professionals alike due to its efficiency and the increasing availability of inexpensive handheld receivers that can calculate a position based on satellite signals (Chapter 8).

The UTM/UPS system has several significant advantages over latitude and longitude. First, unlike the latitude and longitude coordinate system where the distance covered by one degree of longitude increases toward the poles, UTM/UPS coordinates are based on a grid at a constant scale. Second, coordinates are measured in metric units and are decimal based so they are easy to calculate and there is no conversion necessary. Among its other advantages, the UTM/UPS grid has been designed to exclude negative numbers and south and east designators; grid values always increase from left to right and bottom to top. This was accomplished by assigning a central meridian for each zone an arbitrary value of 500,000 meters easting, and fixing a value of 10,000,000 meters northing to the equator in the southern hemisphere (pp. 56–57).

If you want to follow-up and get more information about the UTM/UPS system, the USGS Fact Sheet 077-01 *The Universal Transverse Mercator (UTM) Grid* is available on-line at http://erg.usgs.gov/isb/pubs/factsheets/fs07701.pdf. Technical detail on the geometric calculations performed is available in the National Imagery and Mapping Agency (NIMA) publication *The Universal Grids: Universal Transverse Mercator (UTM) and Universal Polar Stereographic (UPS)* is on the web at http://earth-info.nga.mil/GandG/publications/tm8358.2/TM8358_2.pdf.

Latitude and Longitude

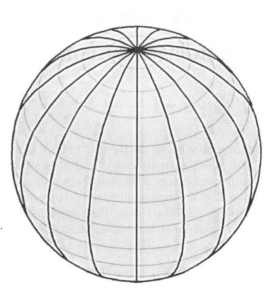

Meridians of Longitude

Lines of longitude are also called meridians. Longitude measures distance west or east of a prime meridian. There are 180° of longitude in each hemisphere, eastern and western. Longitude ranges from 0° at the prime meridian to 180° east and west, on the opposite sides of the globe.

Parallels of Latitude

The equator is a line of equidistance between the poles. Latitude measures distance north or south of the equator. Lines of latitude are also called parallels. There are 90° of latitude in each hemisphere, northern and southern. Latitude is 0° at the equator and 90° at each of the poles. The North Pole is 90°N and the South Pole 90°S.

The Measure of Space and Time

The 0° prime meridian passes through the site of the original Royal Observatory in Greenwich, outside London, also the locus of *Greenwich Mean Time*. On the opposite side of the earth, 180° corresponds to the mid-Pacific *International Date Line*.

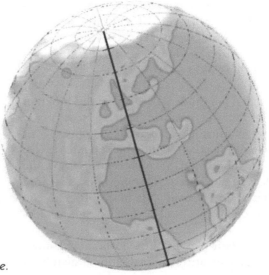

Why are quad sheets also called "seven-and-a-half-minute sheets"?

Quad sheets are 7'30" (7.5 minutes) of longitude and 7'30" (7.5 minutes) of latitude.

Latitude and Longitude Definitions

Degrees, Minutes, Seconds

Precise positions are logged using the full phrase:

122°07'30"W
40°22'30"N

Degrees   Minutes   Seconds   Hemispheric Position

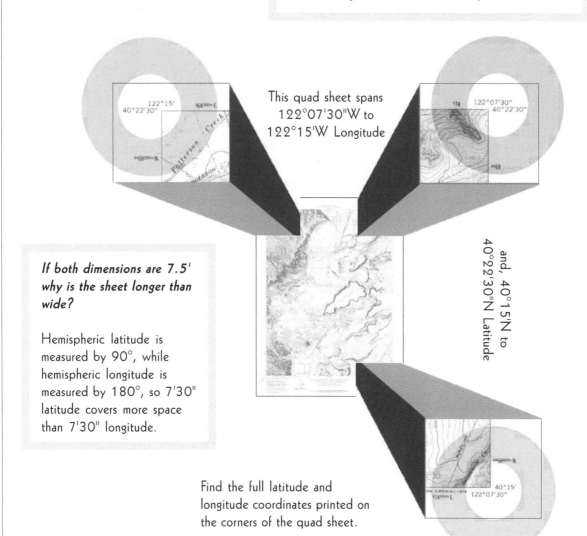

This quad sheet spans 122°07'30"W to 122°15'W Longitude

and, 40°15'N to 40°22'30"N Latitude

If both dimensions are 7.5' why is the sheet longer than wide?

Hemispheric latitude is measured by 90°, while hemispheric longitude is measured by 180°, so 7'30" latitude covers more space than 7'30" longitude.

Find the full latitude and longitude coordinates printed on the corners of the quad sheet.

Latitude and Longitude Plotting and Recording I

First: Note the corner positions of the map. The map lists the precise lat/long location of each corner. In the upper right and left corners, longitude is above, aligned with the left and right borders, and latitude below, aligned with the top and bottom borders.

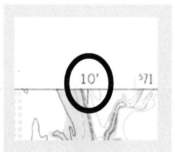

Second: Find the tick marks. Latitude and longitude increments are registered by fine black tick marks aligned inside the map frame at 2'30" intervals around the margins of the map.

Third: Establish measurement scales for latitude and longitude. Measurement will require conversion because quad sheets provide no fixed lat/long scale. Further, in mid-latitudes the lat/long scales are different *on the same map*, requiring different conversions for each. The conversions are made easier if you have access to a surveyor's 4-way or 6-way map scale. If not, any scale will work but using a metric or other scale in units of ten makes the calculation easier. Here's how to figure longitude:

Measure from the map corner to the nearest 2'30" tick:

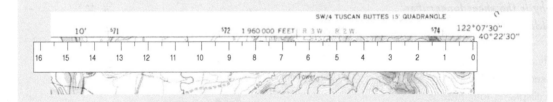

In this case, 2'30" (150") is 14.6 cm.
Next, divide the seconds by the measure, or 150"/14.6 cm=10.3.
Repeat the exercise for latitude. On this map 150"/19.1 cm=7.8.

So, on this map (at latitude 40°N):
1 cm = 10.3" longitude
1 cm = 07.8" latitude

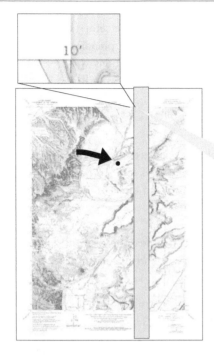

Latitude and Longitude Plotting and Recording II

Fourth: Measure longitude. Find the nearest longitude line east (right) of the subject point, in this case the 10'W tick. Align a straightedge exactly to matching 10'W ticks, visible at the top and bottom of the map. As shown here, make sure you lag the straightedge to the right (east) of the tick marks when you are measuring longitude. That way the straightedge will not block your view of the scale or subject point to be plotted or recorded. Draw a line connecting the ticks or hold the straightedge in place.

Fifth: measure the distance from the tick to the subject point using the metric ruler. In this case, 2.85 cm.

Using the conversion factor derived from step three, convert to seconds: 2.85 cm x 10.3" = 29.35"

Sixth: calculate: Longitude=corner+tick+measure=

120°07'30" + 2'30" + 29.35" = 120°10'29.35"W

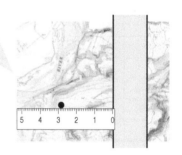

Seventh: repeat the exercise to find latitude. Use a straightedge to find the nearest latitude line south of (below) the subject point. Again, align to matching ticks, in this case the 20'N mark. Lag the straightedge below (south of) the tick marks. Measure, then convert using the latitude rate (07.8" per cm).

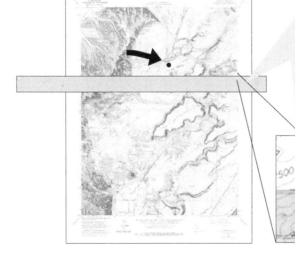

Universal Transverse Mercator Definitions

Universal Transverse Mercator (UTM) coordinates are used to pinpoint and describe horizontal positions with reference to a two-dimensional grid.

The UTM Grid

The (UTM) system divides the world into 60 north-south numeric columns numbered easterly beginning with Zone 1 at the international date line (180°W) and 20 east-west alpha rows lettered northerly beginning with Zone C at 40°N.* Numeric zones are set every 6° longitude and alpha zones every 8° latitude.

Grid squares are identified by longitude number and latitude character, in that order. For example, Lincoln, Nebraska is located in zone "14T." U.S. states are contained between zones 2 and 20 and Q and X.

UTM Zones

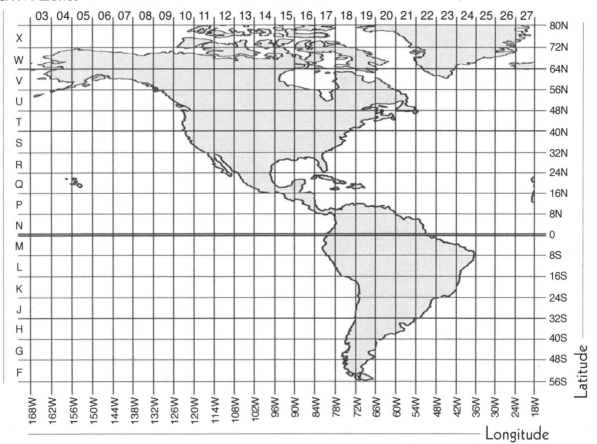

*Alpha zones A, B, Y, and Z are reserved for the Universal Polar Stereographic (UPS) coordinate system, an adjunct to the UTM system adapted to the specific geometric projection required at the poles. O and I are not used in order to avoid transcription errors.

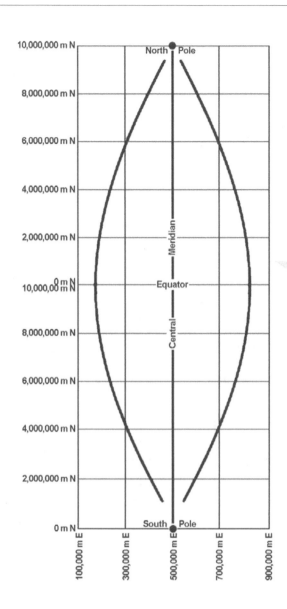

Universal Transverse Mercator Projection and Notations

In order to minimize distortion which would amplify with measurements taken from an east or west meridian, each zone is a Transverse Mercator projection or, more precisely, an equidistant cylindrical projection that "intersects the earth" along a central meridian line. However, all east-west measurements are east of this line, made possible because the arbitrary distance 500,000 m E is assigned to the central meridian. The efficiency of this projection means that distortions of scale are minimal and averaged, ranging from 0.996 along the central meridian to 1.001 along the lateral meridians near the equator.

No Zed, No Westing, No Southing

The "false easting" value of 500,000 meters assigned to the central meridian is ample enough to contain the zone, and thus there are no negative easting numbers. Northern hemisphere positions are measured northward from 0 meters at the equator. For the southern hemisphere, the equator is assigned a "false northing" of 10,000,000 meters and the south pole 0 meters, thus there are no negative or southing coordinates.

Coordinates are listed as "UTM easting" and "UTM northing"

Positions are measured in each zone in meters east relative to a central meridian and meters north relative to the equator. List longitude first (easting), latitude second (northing).

If you record a feature that crosses zone boundaries, avoid mixing coordinates from one zone with coordinates from another. List the coordinates by zone.

Find UTM Coordinates on the Quad Sheet

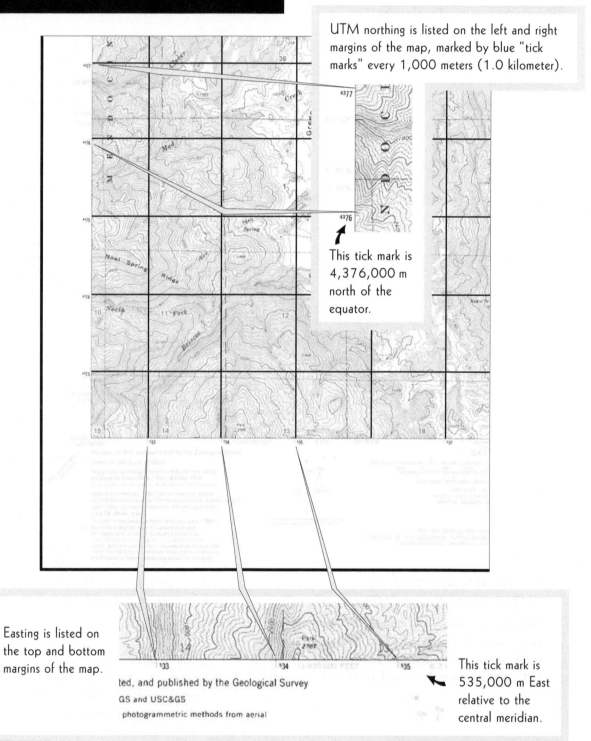

UTM northing is listed on the left and right margins of the map, marked by blue "tick marks" every 1,000 meters (1.0 kilometer).

This tick mark is 4,376,000 m north of the equator.

Easting is listed on the top and bottom margins of the map.

ted, and published by the Geological Survey

GS and USC&GS

photogrammetric methods from aerial

This tick mark is 535,000 m East relative to the central meridian.

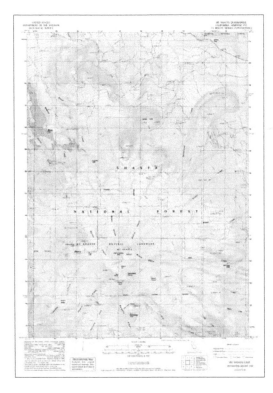

UTM Scale and Template

Recent and military-issue USGS quads have a 1,000 meter UTM grid imprint. Following the military custom, these 1,000-meter grid lines are called "clicks." An objective at five clicks is 5,000 meters away.

Older maps generally do not have the imprint. If not, use a long straight edge to match lines and draw the grid directly onto your field maps.

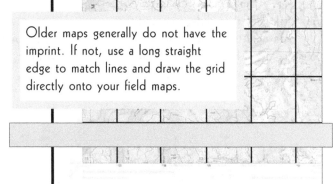

You can make your own UTM scale template. Load a photocopier with translucent acetate sheets (overlay sheets). Set the copier on 100%, photocopy this page, and cut out the template.

Note that this template is setup for 7.5' (1:24,000) maps. If you plan to use 15' maps, set the photocopier for 50% reduction. Everything stays the same.

Check your UTM scale template against the meter or kilometer scale at the base of your quad sheet—make sure it is precisely to scale!

The Archaeological Survey Manual

UTM Land Area (meters)

UTM Scale (meters)

Using the UTM Scale

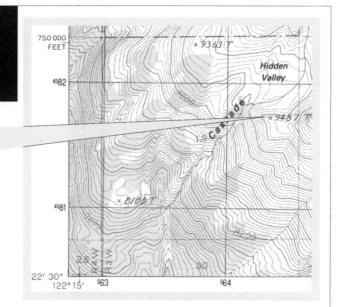

First: Identify a feature and determine its click address. For example, this peak adjacent to Hidden Valley is located in Zone 10T, click 56500 E, 4582000 N.

Second: Align the scale template to the click grid, then slide it up and to the right until the upper right (0/0) scale corner is fixed over the feature.

Third: Add the horizontal distance to the easting interval. Add the vertical reading to the northing interval. *This point is 564360 easting, 4581720 northing.*

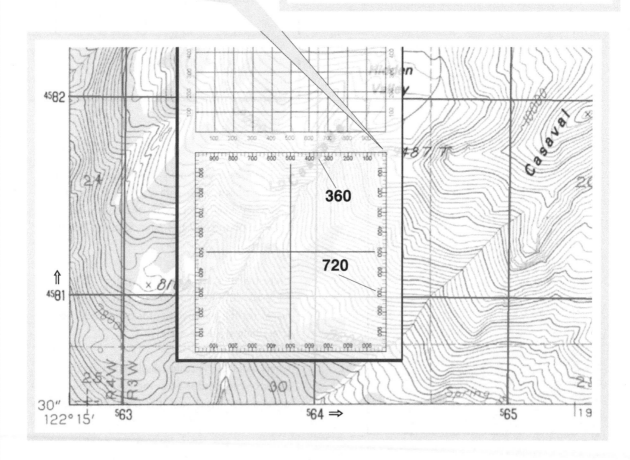

GPS in the Field

Chapter 8

The U.S. Department of Defense manages the *Navigation Satellite Timing and Ranging* system (NAVSTAR), a network of 24 primary and additional backup satellites that broadcast Global Positioning System (GPS) timing signals worldwide. These satellites operate on six orbital planes 20,000 km above the earth's surface in 12-hour orbital periods. Five monitoring stations passively track the satellites and a master control station sends navigation commands and signal uploads containing timing code. The system is designed to keep at least six satellites in line-of-sight from any point on earth at any time.

Research and early development began in the 1950s, and satellites that broadcast navigation timing signals were placed in orbit beginning in the 1960s. The first NAVSTAR satellite dedicated to the current system architecture was launched from Vandenberg Air Force Base, CA, on February 22, 1978, one of four GPS satellites launched that year.

Initially, the system was restricted to military applications. However, increasing demand for civilian aviation and marine shipping use was answered in 1983, when Korean Airlines Flight 007 was shot down by Soviet MIGs after straying over U.S.S.R. territorial waters. The incident was thought to have been the result of the limitations of earthbound navigation signals and this led U.S. President Ronald Reagan to sign an executive order expanding civilian aviation use of the NAVSTAR system, leading to the development of civilian GPS signals equal to the military in accuracy.

The benefits of the NAVSTAR system were made clear in 1991 with the first Gulf War, where GPS navigation proved critical to ground forces in the featureless desert.

The civilian GPS application was fully operational in 1995, and in 1996, President William Clinton formed the Interagency GPS Executive Board to oversee modernization and expanded public and private use. In 2000, President Clinton also signed an executive order terminating *selective availability*, which was a military program that until then had purposely degraded the civilian GPS signals in order to interrupt real-time access to high-accuracy signals on the part of everyday users.

Because the signals and technical specifications are free to public and private users, development of the NAVSTAR system has promulgated a wide variety of civilian products and services, and civilian applications now dominate. Anglers, hunters, recreational hikers, geocache fanatics, and mountain-bike enthusiasts all rely on GPS, and even autos with on-board navigation and combines and tractors in the fields depend on GPS signals. GPS is important in government and scientific enterprises as well, such as search and rescue, weather tracking, habitat mapping, fault and ice sheet monitoring, and ocean navigation.

There have been four generations of NAVSTAR satellites and planning for the fifth is underway, with upgrades to include new military codes as well as improved civilian signal quality and master control efficiency with the ultimate goal of enhanced accuracy using today's receivers.

How Ranging Works

Each satellite broadcasts distinctive time-coded signals of two types, coarse acquisition signals and precise ranging signals, some assigned to military and some to civilian use.

The term "ranging" applies to the use of a receiver and describes the act of determining a position based on acquisition of time signals. The handheld GPS receiver acquires a signal from each satellite, identifies the satellite and its track, deciphers the signal code to determine the difference between time of broadcast and time of reception, and then uses this to calculate the distance between the satellite and receiver [difference in seconds x 186,000 miles per second]. The calculation requires highly accurate clocks, and the NAVSTAR satellites are equipped with on-board atomic timers.

By calculating distance from four or more satellites, the receiver can determine a position. A position is calculated in three dimensions—four counting time, which tends to be the most accurate or finitely calculated dimension. Velocity is also part of the calculation on some applications.

Select a Good Receiver

The first portable GPS receivers were heavy and bulky, with steel armor designed for military use. Today's receivers are light and compact electronic devices designed for belt or pocket attachment.

Because archaeologists operate in a variety of environments—from cities to rugged outlands—where multi-path errors are likely, you should select a minimum 12-channel receiver. The receiver should provide an easy user interface with pages dedicated to tracking, waypoint entry, satellite status and signal strength indicators, and on-board mapping. Make sure your receiver provides a full set of system options including a map datum selection interface, and navigation, coordinate system, and measurement unit options. The receiver should also have the capacity to store a minimum of 50 routes and 1,000 waypoints, track at variable time increments (enabling you to track your survey path), and permit storage of multiple waypoints with alphanumeric user identification options. Many units now come with the capacity to increase memory, but look for adequate on-board memory first. An external antenna attachment may or may not provide a significant boost—check product specifications before you buy.

Modern GPS units are generally designed to easily interface with other electronic devices. Some units rely on parallel port or serial cable connections, but try to find a unit with a USB (Universal Serial Bus) or wireless connector. Look for a unit that permits a variety of download options including access for common Geographic Information System (GIS) or mapping software programs, most of which contain GPS data access routines. You should also consider a GPS unit

that has upload or card-slot options for maps, coordinates, or other data sources.

Four things to really watch for—and these make a real difference in the ease and efficiency of your field work. First, make sure the user interface screen is well-lit (if full-color) or has contrast adjustment (if LCD). Some screens "washout" in full sunlight, a problem generally solved by increasing contrast. Second, make sure the unit has a battery-status indicator so you can change low batteries before you go in the field. Third, try to find a unit that is good with energy conservation. Otherwise, you may find yourself changing batteries two or three times per week, an expensive turn of events. Finally, try to find a waterproof GPS unit. These things pop out of packets, belt packs, and cradles at the most inappropriate times!

Advantages of the NAVSTAR GPS system are many and varied, but for the archaeologist engaged in conventional archaeological survey, a good, off-the-shelf GPS receiver will supply:

ï 24-hour service
ï accurate X-Y-Z location information
ï Precise timing
ï unlimited worldwide use
ï location in a systematic grid

For More Information

For more information on the NAVSTAR system, visit the *NAVSTAR Global Positioning System Joint Program Office* at http://gps.losangeles.af.mil/, the *National Space-Based Positioning, Navigation, and Timing (PNT) Executive Committee* website at http://pnt.gov/, or the *U.S. Department of Commerce National Geodetic Survey* home page at http://www.ngs.noaa.gov/. Aeronautical applications are discussed on the *Federal Aviation Administration Satellite Navigation* page at http://gps.faa.gov/, and nautical applications are covered in the *U.S. Coast Guard Navigation Center GPS* home page at http://www.navcen.uscg.gov/gps/default.htm.

The *Aero.org GPS Primer* at http://www.aero.org/education/primers/gps/index.html is a very solid overview of the system. An interesting overview of plans for the fifth generation of NAVSTAR satellites is presented on the *Boeing Integrated Defense Systems GPS IIF/III* web page at http://www.boeing.com/defense-space/space/gps/.

GPS Ranging

The term "ranging" refers to the act of a user determining position based on acquisition of a satellite-based time signal.

The GPS receiver determines distance to each satellite and calculates position as the location where all vectors intersect.

Access to signals from four or more satellites allows the receiver to calculate a specific position. A 12-Channel receiver can receive signals from 12 satellites simultaneously.

With just two satellites, user position could be anywhere along the perimeter formed by the intersection of two signal gradients.

Addition of a third satellite signal reduces the alternative positions to one of two points of intersection.

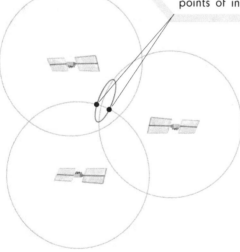

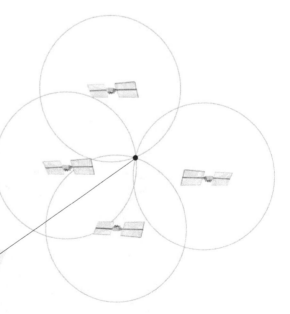

A fourth satellite signal fixes position to a single point of intersection.

Set a Common Datum

As you prepare for the field, make sure your GPS unit is set to the same datum as the quad maps covering your project area.

An incorrect datum match can cause plotting and navigation errors anywhere from a few meters to several hundred meters.

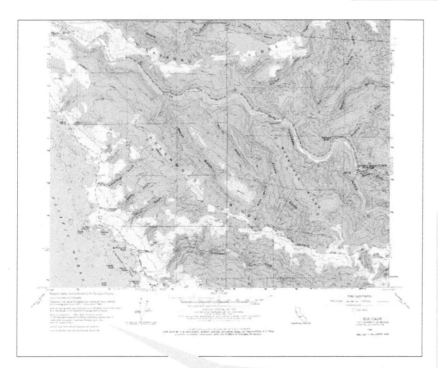

Set your receiver datum in the system preferences interface. Scroll through until you find a match.

┌**System**
└─**Navigation**
　└─**Datum**

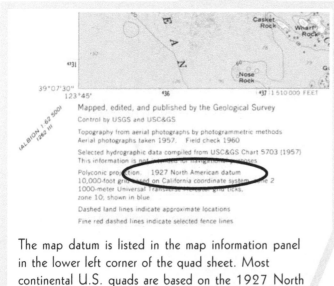

Mapped, edited, and published by the Geological Survey

Control by USGS and USC&GS

Topography from aerial photographs by photogrammetric methods
Aerial photographs taken 1957.   Field check 1960

Selected hydrographic data compiled from USC&GS Chart 5703 (1957)
This information is not intended for navigational purposes

Polyconic projection.   1927 North American datum
10,000-foot grid based on California coordinate system, zone 2
1000-meter Universal Transverse Mercator grid ticks,
zone 10; shown in blue

Dashed land lines indicate approximate locations

Fine red dashed lines indicate selected fence lines

Select Datum

NAD27 Greenland
NAD27 Cuba
NAD27 CONUS
NAD27 Central
NAD27 Caribbean
NAD 27 Canal Zone

The map datum is listed in the map information panel in the lower left corner of the quad sheet. Most continental U.S. quads are based on the 1927 North American Datum (NAD27 CONUS) or the 1983 North American Datum (NAD83 CONUS).

Set Coordinate and Measurement Systems

Make sure your GPS system preferences are set to the same coordinate system you plan to use in other forms of documentation as you navigate, track, record, and plot archaeological observations.

The Universal Transverse Mercator (UTM)/ Universal Polar Stereographic (UPS) coordinate system is in common use worldwide and in the sciences.

If you use UTM/UPS coordinates you might also change your measurement unit to the metric system. This will help you maintain systematic documentation and produce ranging and tracking records that can be efficiently plotted on quad sheets with the click imprint.

However, if you do this, most units will not only change the distance unit but also the elevation unit and therefore give you elevation in metric. If you are unused to this way of thinking about elevation or your quad sheet shows elevation in feet, then there is no harm in keeping this setting on statute.

Of course, all these provisions may be unnecessary if you are following historical documents recorded in latitude and longitude. In this case you should consider using one unit to track (in lat/long) and one to record (in UTM/UPS).

Be Aware of Potential Sources of Ranging Error

User errors are common—most related to poor reception, for example, errors that result from registering a position before the receiver has acquired sufficient satellites with adequate signal strength, or reliance on poor signals obscured by a thick canopy.

Multipath errors caused by signal reflection are common in mountainous terrain, canyonlands, and cityscapes. This can be solved by using strong antennas and multichannel receivers capable of making the best adjustment.

Give the receiver enough time to acquire four or more satellite signals at optimum signal strength. You may have to return and try again.

Field Work

Chapter

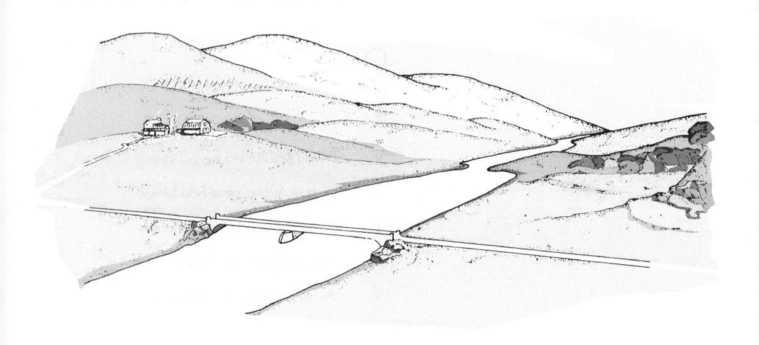

Griffin Valley — the hypothetical example used in the
following chapters — as it appears today, looking northeast.
That's the Ford Farm on the hill above the Phillips River.

Prefield Research and Survey Design

Chapter 9

Students, managers, and cohorts in allied professions frequently express frustration at the pervasive "gray areas" that muddy the archaeological enterprise. For example, archaeological survey is not one thing governed by uniform guidelines and specifications. There has never been a detailed definition of the term "archaeological survey" on which all archaeologists—and others who use the word—have agreed. There are many standards used nationwide and even within federal jurisdiction there are agency-by-agency differences. Consequently, when a qualified archaeologist performs a survey we can't assume that it will be conducted to a particular professional standard. What in one context might be call a "survey," in another might be labelled a mere cursory inspection. Moreover, as noted in Chapter 2, the reasons for and methods of making surveys have changed considerably during the last century and a half and, most dramatically, during the last three decades. As a result, it is entirely predictable that if an area surveyed 20 years ago were surveyed again today many archaeological sites not noted in the first survey would be newly discovered on the second pass.

Thus, from the standpoint of actions and outcomes, chaos appears to reign. However, there is no reason to lose hope; this apparent chaos is actually welcome and a sure sign of healthy, problem-oriented field work. Archaeologists are not guided by one single-minded purpose—like finding all sites, or the biggest, or the oldest, or best-preserved—but by many and varied purposes driven by project-specific management and research needs. Thus, in its brightest manifestations archaeological survey design is guided by the prior articulation of its purpose, whether management, research, or—more often the case—both. Specific problems should be identified and field tasks and methods designed to solve them. Given thoughtful planning and adequate resources it may be possible to identify as much as one needs to address the survey's purpose. In other words, archaeologists should control their methods, not the other way around.

This chapter explores the relationship between survey purpose and survey design and the predictable pitfalls to avoid in planning an archaeological survey. In order to describe the variety of archaeological survey types and activities and their results in the context of a standard environmental setting, we present Griffin Valley, in the state of Indeterminate.

THE GRIFFIN VALLEY ARCHAEOLOGICAL RECORD

The Human Past in Griffin Valley

Griffin Valley lies in relatively gentle, rolling country with a good deal of environmental diversity, along the Phillips River. Much of the valley is a part of the Ford Ranch, whose 75-year-old buildings appear toward the left side of the illustration shown on the facing page. Excluding these buildings from consideration for the moment, we can define the nature of the valley's archaeological resources for the purposes of our example.

Human beings first entered Griffin Valley about 11,000 years ago. At that time, toward the end of the Pleistocene ("Ice Age"), much of the valley was covered by a pluvial lake (p. 70, top). The lake was shallow and marshy, and many large herd animals came there to drink. Waterfowl abounded. Because it was an ideal place for hunters to live, a small wandering band established a campsite at the low pass near the future location of the Ford Ranch buildings. These people produced and used what are now referred to as "Clovis points"—distinctive flaked-stone spearpoints with channels flaked into either side to accommodate their attachment to shafts.

The spot the Clovis people selected for their camp afforded them some shelter from the elements and was close enough to water to be convenient but not so near

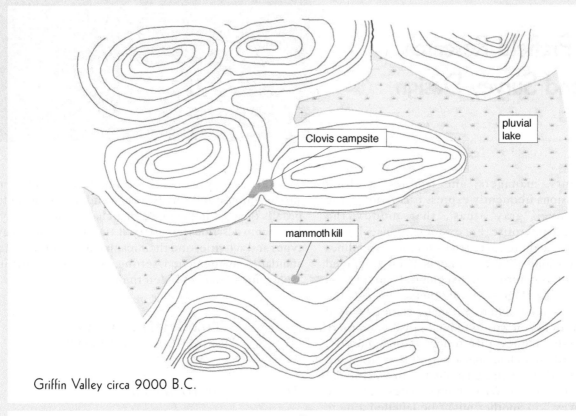

Griffin Valley circa 9000 B.C.

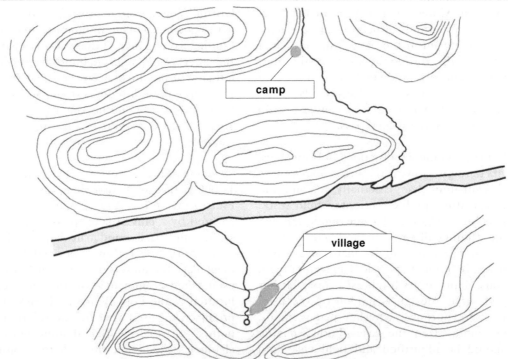

Griffin Valley circa 2000 B.C.

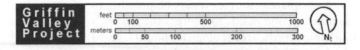

as to frighten game away from the shore. It commanded a view of both the lake and the small valley to the north, down which game often passed. Generations of band members, who ranged seasonally over a large territory, visited this site recurrently for several centuries and hunted around the lake margins. One season, three hunters from the group surprised a mammoth foraging along the south shore of the lake. Floundering around, the mammoth became mired and could not escape. The hunters waited for him to weary, and then dispatched him with many spears. The entire band then moved to the kill site and butchered the beast, leaving his bones, most of the spearpoints that had killed him, some of their butchering tools, and their firepits when they moved on.

As the glaciers retreated at the end of the Pleistocene, a more diversified sort of hunting and gathering came to dominate the human economy of the area. Now the lake was gone and grasslands covered much of the valley. The really big game was also gone, and vegetable foods played a larger role in the diet of local people. Small, hard seeds from grasses were ground on milling stones, and small game was hunted. During this period a good spring flowed out of the low rocky mountains at the south side of the valley, and it was around this spring that a good-sized semi-permanent village was established, (p. 70, bottom). This was a very convenient location, with easy access to fresh water and grasslands, and a short walk from the sage-covered low hills where the hunters had camped three thousand years before. During one period of about a century the climate turned arid and the available seed crop grew very sparse. Women now had to range farther afield to gather an adequate supply of seeds. A temporary overnight camp was established at the north edge of the valley, near a creek at the edge of the sage fields. Here seeds could be stockpiled and ground before being transported back to the main village; men accompanying the women could hunt in the nearby chaparral.

About 3000 years ago, a violent earthquake sealed up the spring, and the villagers had to move. Their new settlement was located at the foot of the pass through the Ford Ranch hills, on the bank of the Phillips River near the ecotone between grassland and chaparral communities. The oaks on the north slope of the hills were within easy reach, which was good since the people had recently developed techniques for leaching the tannic acid from acorn meal and making it edible. With this new source of food, and a moderating climate, the population increased rapidly and was soon in

danger of exceeding the carrying capacity of the local environment. Fortunately, at this juncture, some of the people's trading partners to the south introduced them to maize, and soon they had learned to plant and grow this important crop along with beans, squash, and sunflowers. At first, crops were planted along the floodplain at the immediate margins of the river, but later gardens were extended farther across the plain to the south (p. 72, top).

Population was now increasing elsewhere and strife inevitably followed as different groups sought to expand their territories. After being virtually wiped out twice by neighboring groups seeking their food supplies, the people of Griffin Valley reluctantly relocated their village to a less convenient but more defensible site: the crest of the ridge of hills east of the pass. Here they built a strong palisaded village. New fields were established along the north side of the hills, and the small creek was diverted to irrigate them.

In these times of stress, a religion developed that centered on arduous male initiation rites. Such rites prepared 10 to 12-year-old boys for the rigorous, dangerous lives they would lead as men. At one point in the ritual, each boy was required to run silently to the crest of the mountains to the south, where his tutor (usually his mother's brother) awaited him. The tutor helped the boy assume a difficult position under one of the many overhanging rocks that topped the mountains, bending over backward with his nose a few inches from the top of the overhang. With a hammerstone, the boy was then required to peck a small, cup-shaped depression in the roof of the overhang. The work had to be done in silence and without food; it typically took 2 to 3 days, during which time the boy's tutor instructed him in the history and ethics of the tribe and discussed what it meant to be a man. By the time the ordeal was over the boy was usually hallucinating; he was given paints and encouraged to illustrate his visions on any rock of his choosing.

In A.D. 1710, a French trapper brought the people their first iron tools and glass beads. In 1778 they were attacked, and their village was burned by a group of Seneca fleeing the decimation of their own homes by Continental troops. In 1780 a smallpox epidemic swept the community, leaving many dead. By this time the great palisaded village was no longer needed, and after the Seneca attack it had never been effectively rebuilt. The people now took up residence near their irrigated fields at the north edge of the valley (p. 72, bottom).

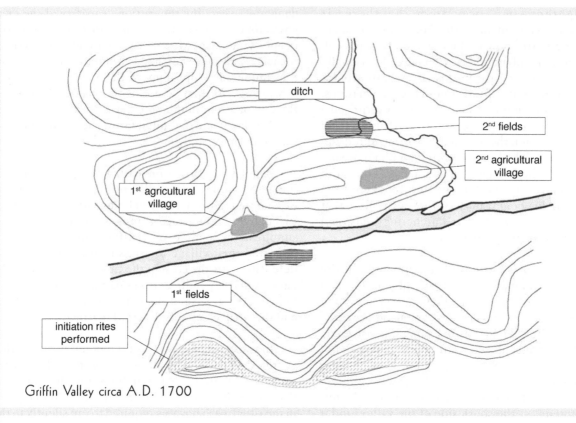

ditch

2nd fields

2nd agricultural village

1st agricultural village

1st fields

initiation rites performed

Griffin Valley circa A.D. 1700

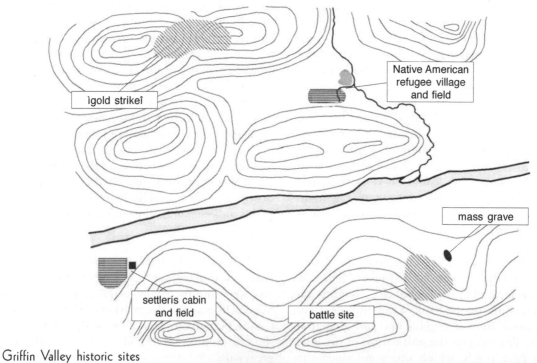

ìgold strikeî

Native American refugee village and field

mass grave

settlerís cabin and field

battle site

Griffin Valley historic sites

Griffin Valley Project

feet
0   100   500   1000

meters
0   50   100   200   300

N↑

In 1820 the first white settler arrived, built a cabin, and established a small farm on the south bank of the river. By 1850 the white population in the area was substantial, and settlers began to worry about the threat posed by the Indians. They petitioned the U.S. government to rid them of the Indian peril, whereupon the government obliged by creating a reservation to which the various scattered tribes would be relocated. Because the refugee occupants of Griffin Valley did not want to go, they were removed by force. Although one group broke away and fortified an area in the rocky slope south of the valley, they were promptly and easily overwhelmed by a troop of irregulars from the nearby town, massacred, and interred in a common grave.

Once again the valley lay uninhabited. The Indians had been removed or murdered and the valley's one white settler had abandoned his farm and fled to town during the period of unrest. It became part of a large and informally bounded cattle ranch, and no one lived there for a number of years.

In 1872, a wandering miner reported finding gold in the mountains north of the valley. More than 5,000 would-be millionaires descended upon the scene of the strike, only to discover after less than a month that the gold discovery had been a hoax to divert attention from a real strike about 100 miles away. The site was immediately abandoned and promptly forgotten (p. 72, bottom). In 1890 B. J. Griffin established a cattle ranch in the valley and in 1895 sold out to A. R. Ford, who in 1890 built the house and barns that remain the ranch center today.

What the Past Has Left Us

Eleven thousand years of human history in Griffin Valley have thus created a rich mosaic of archaeological sites (below)—the physical expressions of the many things people have done there over the centuries. None of these sites has a neon sign on it saying "Archaeological Site," or "Dig Here." In order for anyone to know about them, they have to be found.

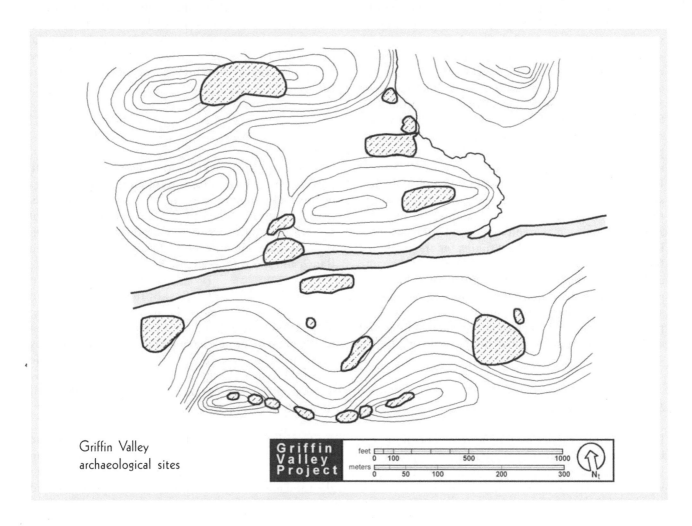

Griffin Valley archaeological sites

Finding them would be easy if in real life we knew all the things about the valley that have been recounted above and knew that our knowledge was accurate and complete. But of course, we never start out knowing these things. We have to learn them, and a major source of information on what happened in a place like Griffin Valley is the archaeological record—the very sites created by centuries of human life in the area. So if we are going to learn about Griffin Valley's human past, we have to discover its archaeological sites. This, of course, is why we do archaeological surveys—if we are archaeologists, or other kinds of people interested in the past.

PREVIOUS INVESTIGATIONS

Why a Non-Archaeologist Might Require a Survey of Griffin Valley

A manager—whether managing land, a government agency, or a project of some kind—may or may not be interested in Griffin Valley's past, but depending on what she is managing and the legal context in which she is managing it, she may need to undertake, oversee, or at least finance archaeological surveys. For example, one of the basic responsibilities of the Indeterminate State Historic Preservation Officer, as outlined in Section 101(b)(3) of the National Historic Preservation Act, is to develop a statewide inventory of "historic properties"—that is, districts, sites, buildings, structures, and objects eligible for the National Register of Historic Places—including archaeological sites. And the manager of a project that would modify the valley in some way—for example, by building something in it—may need to be involved in a survey if his project involves federal funds, federal land, or a federal permit, triggering the requirements of the same act's Section 106. Section 106 regulations, while they do not necessarily require the conduct of surveys per se, do require that historic properties affected by a project be identified for consideration in planning, and such identification may not be possible without a survey.

Surveying Griffin Valley

Let us assume, then, that for one reason or another—because we are conducting the state historic properties inventory, because we are planning to fill the valley with toxic wastes and need a federal permit to do so, or

because we are simply interested in the area's past—we want or need to do a comprehensive survey of Griffin Valley's archaeology. How will we do it?

We'll undoubtedly start out by ascertaining, as best we can, what we already know, or think we know, about the valley and the larger area of which it is a part. By "we," I mean not only ourselves personally but the world as a whole, or at least the community of scholars who have written about the past. Once we have our existing information under control, we'll decide whether it's sufficient for our planning purposes—whether it gives us a sufficient understanding of the valley's archaeological sites to plan for their management. If not, we'll need to figure out what further, new research is needed.

Review of Previous Survey Results

So, what do we already know about Griffin Valley? To find out, most archaeologists would turn first (and quite often only) to the official files maintained by the area's archaeological authorities—such as a state historic preservation officer or a university's archaeological program. In our case, the Indeterminate State archaeological site files include record of two previous surveys.

Dr. Beakey's Notes

In 1932, a group of deer hunters discovered elaborate polychrome paintings on certain protected overhangs in the rocky mountain south of the valley. In 1938, local enthusiasts persuaded Dr. Linford Beakey of Indeterminate University to view the pictographs. Dr. Beakey, at the direction of the locals, drove to the Ford Ranch and was warmly greeted.

Mr. Ford showed Dr. Beakey the flint tools he had found in the field below the house. The party then walked down the hill, and Mr. Ford pointed out the spot from which the projectile points had come. Dr. Beakey noted what he described in his notes as a probable extensive Late Stoneland village site, represented by discolored soil, fire-fractured rocks, and the flakes of stone left over from tool and weapon production.

Crossing the river at a low point slightly west of the site, the party crossed to the toe of the hills and began to climb. Although Dr. Beakey noticed a scatter of old glass and metal fragments at the foot of the hills, he did not make any record of this in his notebook; as a student of prehistoric Native American cultures, he had neither interest nor competence in the study of historic sites.

At the crest of the ridge, Beakey photographed and sketched several polychrome rock faces, and pointed out to his associates that cup-shaped petroglyphs were also to be found on several overhangs; they had not noted these rather nondescript features and were not especially impressed.

After an hour or so spent inspecting the area, and eating an excellent picnic lunch, Dr. Beakey returned to the ranch house via the canyon east of the slope up which he had climbed (p. 76, top). While so doing, with the sun's rays now slanting in low from the west, he noticed irregularities in the contours of the floodplain. These might be field scars associated with the village site which had earlier attracted his attention. He made notes concerning this possibility, but could not see the ridges once he reached the floodplain. While looking for the field remnants, he failed to notice the bone fragments in the dirt that gophers had pushed out of their holes at the edge of the oak woods.

The above is what actually happened, but Dr. Beakey's notes don't report it all, and don't report what he failed to do or see. Finished with the day's work, Dr. Beakey filled in one of his university's one-page site-survey forms describing the village site he had recorded and another on the several examples of rock art. He placed these, together with his two notebook pages of notes about these sites and the possible aboriginal field, in the university's archaeological survey files.

The files thus recorded that at least two prehistoric sites existed in Griffin Valley—a Late Stoneland village site and an unknown number of polychrome pictographs and cupule petroglyphs—and the possibility of some ancient field scars. The files did not record the presence of the settler's cabin site, which Beakey had noticed but had not found interesting enough to record; nor did they record the mammoth kill site which he had failed to note while concentrating on something else.

Nor did they indicate the portions of the valley and its surrounding hills that he had inspected and those that he had not. They did not indicate how he had inspected the area, and they did not indicate what he

had been looking for. From the files alone, it would be impossible to determine whether Dr. Beakey had simply walked across the valley on a sunny day in the pleasant company of some local hunters and ranchers—as indeed he had—or whether he had spent a month crawling over it on his hands and knees. Compared with the written results of many archaeological surveys done in the early to mid-twentieth century, however, Beakey's notes were pretty detailed.

The Highway Survey

In 1965, the Indeterminate Highway Department (IHD) proposed construction of an expressway through the valley from southeast to northwest. The project was eventually abandoned after a prolonged lawsuit filed by local landowners and environmentalists, but during planning for the project an archaeological survey was conducted. While the enlightened and progressive IHD was unusual among highway agencies of the day in that it was willing to pay for surveys, it placed restrictions on its survey parties; they were not permitted to range beyond the already selected highway right-of-way, and no funds were provided for either preliminary background research or post-survey analysis and report preparation. Those few highway departments that funded archaeological surveys in the 1950s and 60s commonly imposed such restrictions.

Constrained by a modest $500 that had been allotted for the survey—also typical of the period—only a brief surface inspection could be made. The Indeterminate University Anthropology Department detailed a young archaeology graduate student, E. M. Loumington, to conduct the survey. Loumington began by reviewing the university's site files, and hence looked over Beakey's notes, but had neither the time nor funds to perform further background research. She then undertook a field survey of the right-of-way, carefully avoiding contact with the irate local residents (p. 76, bottom).

Hiking down the southern mountains from the southeast, Loumington noted a peculiar, generally rectangular depression in the ground at the foot of the rocky slope. A very old, rusty and broken shovel blade was the only cultural evidence she noted in the vicinity of this depression. Concluding that it was a topsoil source used by the ranchers at some time, she proceeded with her survey without recording the discovery.

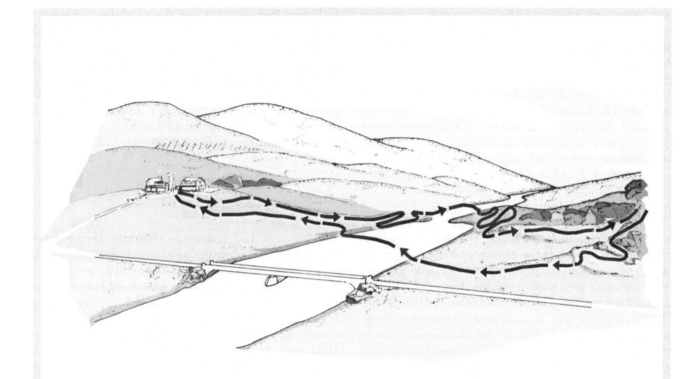

Beakey survey, 1938

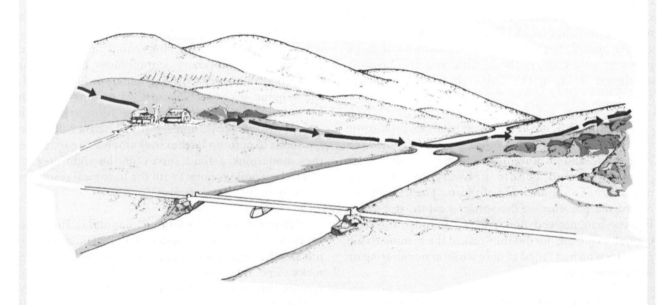

Loumington survey, 1965

Although she crossed the river east of the Late Stoneland site that Beakey had recorded and within clear view of it, Loumington disregarded the site because it was outside the highway right-of-way. Reaching the crest of the ridge at about 11:30 am, she looked into the valley to see if Beakey's field scars were still visible. She discerned no evidence of cultivation because, in the light of the midday sun, there were no telltale shadows to reveal them.

Continuing on, she crossed the north arm of the valley and climbed the northern mountains, noticing a light scatter of old tin cans and bottles but nothing of a prehistoric nature. Having been trained, like most North American archaeologists, to equate archaeology with prehistory, she ignored the historic trash, and recorded none of it.

Thus, Loumington's survey recorded nothing in Griffin Valley. Although she crossed the mass grave of the massacred Indians, she did not recognize it because, having done no background research besides a records check in the university's site files, she did not know that such a thing might be expected. Restricted to the highway right-of-way, she passed close by the large palisaded village without recording anything about it. Because of the time of day, she was unable to see the field scars. Her training as a prehistorian biased her against recognizing the remains of the abortive 1872 gold rush. Her report to the IHD indicated that the right-of-way had been surveyed, with negative results; maps and field notes were filed with the Indeterminate University Archaeological Survey.

The Limitations of "Known Site" Data

So, all we know based on records now in hand is that two surveys have been done in Griffin Valley, one of which revealed the existence of a single Late Stoneland village site and some rock art, while the other revealed nothing. We have some idea of where Loumington looked and the methods she used; we have no record of where Beakey went, other than that he obviously visited the sites he recorded.

This information is clearly insufficient as a basis for understanding Griffin Valley's past. It is worse than insufficient; it is patently misleading. If we accepted it at face value as a comprehensive record of the valley's archaeology, we would miss the Pleistocene sites, the massacre site, the cabin site, and the remnants of the gold rush. But the Indeterminate site records are typical of the kind of information that is available for many portions of the country today.

For this reason, existing inventories of archaeological information in university, museum, organizational, or state files almost *never* provides an adequate basis for planning. For this reason, it is never appropriate to consider only *known* archaeological sites when planning, say, a development project, unless one can be sure that the area has been studied very thoroughly, very comprehensively, and that there are few if any *unknown* sites lurking behind the hills or down in the soil strata.

Recognizing the inadequacy of the existing site inventory data, we decide we need to perform further survey. If we're wise, however, we won't immediately head for the field. There are other sources to be inspected.

The Larger Framework

When the highway project's opponents were successful in keeping it out of Griffin Valley, the highway was routed through Ritchie Canyon, several miles to the west. Development followed the highway, and Griffin Valley remained in agricultural use. When environmental and historic preservation laws were enacted and began to be implemented in the 1970s, archaeological surveys were done in advance of some development projects along the highway, but because no destructive government projects were planned in Griffin Valley, archaeologists have not done any work in the valley since Loumington's survey.

But archaeologists have conducted surveys and excavations in the general area—along that highway in Ritchie Canyon, thirty miles to the east in the valley flooded in 1965 behind Jennings Dam, and over in Ben Wheat Valley, where Gimsey MacRoberts did his dissertation research. And archaeologists aren't the only ones who have gathered information relevant to Griffin Valley's past. There have been studies of the area's Pleistocene geology and geomorphology that have hinted at the one-time existence of a lake in the valley. Paleontologists have found mammoth and mastodon bones in swamps three counties away. Cultural anthropologists have documented the ways of life and religious beliefs of the Ebirt Tribe, which occupied the general area until it was decimated by disease and warfare in the 17th and 18th centuries and marched off to a reservation in Oklahoma along what is called the Road of Weeping. The Ebirt themselves have recently formed a cultural committee and established a tribal historic preservation office that is compiling data on the tribe's aboriginal territory. Professor Patricia Placer at Indeterminate State University's history department has made herself the world's leading authority on the 1871–3 Indeterminate gold discoveries, putative and real. Gloria Griffin has compiled an extensive worldwide web site documenting and illustrating her family's history. Some of these sources of information are published, many are not. More and more of them are being made available on-line. There is a substantial body of information that is relevant to Griffin Valley's past, besides the Beakey and Loumington documents. All this background data can be useful in planning and informing an archaeological survey, and many of those involved in compiling and developing it will be important people to consult as the scope of the survey is developed and research is conducted.

BIAS CONTROL, RESEARCH DESIGN, AND BACKGROUND

Investigator Bias

Certain cinematic exceptions notwithstanding, archaeologists are not super-people. Like any other member of the human race, an archaeologist's perceptions of the universe are influenced by training, interests, and values. If a survey is to be truly comprehensive, the biases of the surveyors must be taken into account and a balance must be maintained.

Some perceptual failings are obvious: one should not employ a color-blind archaeologist to look for polychrome pictographs, and an archaeologist trained only in prehistory should not be expected to locate and identify many kinds of recent historic sites—at least not without assistance.

Other biases are less immediately apparent and may not be recognized, or acknowledged, by the archaeologist. An archaeologist whose experience has been limited to sites that are indicated by surface concentrations of pottery may miss whatever marks the surface of another type of site—for instance, the melted mud-brick walls of a historic stable. He or she must, in a sense, recalibrate his or her perceptions.

Usually, the most sensitive biases—in the sense of being difficult for the archaeologist to recognize and

acknowledge—are those that relate to the expectations and research interests of the archaeologist. I once encountered a case in which highly qualified archaeologists had recorded large numbers of occupation sites without ever noticing associated petroglyphs and pictographs (King 1975: 88). What most likely happened is that the archaeologists whose research interests involved the study of culture-history and environmental adaptation through the excavation of occupation sites, simply kept their eyes focused on the ground rather than on the rock outcrops. Similarly, it used to be quite common for archaeologists whose primary interests were in the reconstruction of culture history through the study of deeply stratified sites to ignore small, shallow, or disturbed sites as "insignificant." Around the time the original version of this volume was published, two colleagues in the Interagency Archaeological Services Division—Valerie Talmadge and Olga Chesler—published a neat little volume demonstrating that such sites often contain important data (Talmadge and Chesler 1977). Many others who have studied such sites have reached similar conclusions. Even if in a given instance they do *not* contain useful information, it is putting the cart before the horse if one fails to record them. Particularly where we're surveying in order to generate management information, we ought to record—at some level of detail—all the archaeological sites that a reasonably efficient search procedure reveals. Once we have a body of reasonably comparable data on all the sites we've found in an area, then we can decide which ones are worth managing and which are not.

One of the most difficult problems facing the manager who contracts for archaeological services lies in recognizing whether the archaeologist has allowed his biases to narrowly determine what will be recorded. For this reason it is essential that archaeologists clearly set forth their assumptions about what makes a site worth recording, for example, before they undertake the survey, and justify each such assumption.

Obviously, there is a point beyond which evidence of past human activity is too inconsequential or ordinary to be noted, or too nebulous to detect without inordinate expenditures of time and money. We can't find every spear point ever thrown at an antelope in Griffin Valley, and we probably don't need to record the pile of disposable diapers we find under a bush. The point at which one perceives something from the past to be unworthy of notice, or impossible to find, varies from archaeologist to archaeologist, team member to team member, and between archaeologists and practitioners of other disciplines. The purpose of bias control is not to eliminate these differences but to understand and control them—to reach a conclusion, understood by all concerned, about what will be regarded as important enough to search for and record, and to specify this conclusion for future reference.

Bias control requires basic self-examination: recognizing one's own interests, training and abilities, and comparing them with those likely to be needed in order to identify the full range of sites likely to be present. Where we recognize "blind spots," survey team composition should be altered to eliminate them. If a reasonable attempt is made at bias control, there is no need to be greatly concerned about finding a single investigator with personal expertise in all anticipated aspects of the survey.

There are two catches, however. First, if we're influenced by unrecognized bias, we by definition don't recognize it. Second, if we're going to compare what we can perceive with what may actually exist, we must have some idea about what sorts of sites the study area is likely to contain.

Treating the second of these problems in detail is one reason we do background research, but both problems can partly be resolved through proper research design and review. Formulating an explicit research design and getting it reviewed by others will go a long way toward controlling bias and ensuring an organized approach to the study.

What Is a Research Design?

A research design is nothing more than a systematically organized project plan, explaining what we intend to do, why we intend to do it that way, and how we plan to carry out our intent. It should articulate:

Where we plan to survey (e.g., "Griffin Valley, in Placid County, state of Indeterminate," usually also specifying latitude and longitude, UTM coordinates, or range, township, and section(s).

Why we plan to survey (e.g., "The survey is one part of an interdisciplinary team study of impacts on cultural resources that may arise from the proposed Griffin Valley Toxic Waste Disposal Project, in connection with preparation of an

environmental assessment (EA) under the National Environmental Policy Act and to support compliance with Section 106 of the National Historic Preservation Act.")

Why we've chosen to survey the area we have (e.g., "The area to be inspected has been identified as the project's area of potential effect for archaeology; it includes all locations where disturbance of the ground is expected either as a direct result of the project's construction and operation, or as an indirect result of the project in the form of induced land development. It is distinguished from the areas of potential effect for visual impacts, auditory impacts, traffic impacts, and impacts on traditional land use, which are dealt with separately in the EA.")

What we plan to look for, and why (e.g., "Background data suggests the possibility of Clovis-period sites, probably associated with hunting, Archaic seed-grinding and processing stations, Late Stoneland village sites and agricultural fields, prehistoric ritual sites with pictographs and petroglyphs..."). If we thoroughly lay out what our background research suggests may be out there to be found, we should be able to judge what we are and are not equipped to find, and either design our survey to find it or be explicit in saying that we're not going to look for it.

How we plan to survey (e.g., "We plan to conduct a non-exclusive comprehensive deployed survey with background research and subsurface testing, in consultation with the Ebirt Tribe and making use of the expertise of the Griffin Valley Archaeological Society and special efforts to relocate and describe sites identified previously by Linford Beakey.")

Why we plan to survey the way we plan to survey (e.g., "This strategy is justified by (a) the likelihood of significant late Pleistocene and later sites in the area, based on background research to date; (b) the stated cultural sensitivity of Griffin Valley to the Ebirt Tribe; (c) the severity of impacts likely if the project is constructed; and, (d) the special knowledge of Griffin Valley possessed by the Griffin Valley Archaeological Society, which is made up of local residents who have conducted avocational surface survey and collected artifacts in the area for several generations.")

Who will do, or at least supervise, the work, and what their qualifications are (e.g., "The survey will be supervised by Dr. Marshall Towne of TROWL, Inc., who holds a Ph.D. in Anthropology from Indeterminate State University and has 15 years of experience conducting archaeological and historic preservation field, laboratory, and archival research in Indeterminate and adjacent states.

What will be produced (e.g., "A complete report detailing all aspects of the survey will be produced upon completion of the fieldwork and any necessary laboratory work and other analyses, and will be provided to reviewers in draft for comment no later than .....")

Of course, a research design can do more, outlining what is already known about the area, articulating special research questions to be addressed, detailing special studies that will be done, explaining how team members will be trained to do their jobs, and so on. Many states, institutions, and organizations have established standards for research designs that should be carefully attended to.

Scoping

Figuring out how we're going to conduct our survey, establishing our research design, and detecting and dealing with biases are all parts of what, in the world of cultural resource management and environmental impact assessment, is often called scoping. Scoping simply means figuring out the scope of whatever enterprise we're about to embark upon—such as a survey—but a critical thing to remember about scoping is that it's public and interactive. We don't hole up in a tower somewhere to write the scope of work for our survey. We consult with others—other archaeologists, other professionals, government officials, Indian tribes and neighborhood groups, property owners and artifact collectors, biologists, and soil scientists. We ask them what they think we ought to do; we share our research design in draft and respect and make the best use possible of their comments. We figure out how our work may dovetail with theirs, or with their interests in the area. We ask them what they know about the area, or think about it, that can help us. Public scoping helps us detect and control bias. It may produce ideas about what to look for and how or where to look for it, and it helps us develop strategies that are tailored to the real, on-the-ground situation. It can also help avoid conflicts with other people and other interests and serve as the basis for fruitful collaboration and cooperation.

In emphasizing the public nature of scoping, we don't mean to imply that everything needs to be revealed to everyone. Indian tribes and other groups often want to keep certain bodies of information secret—notably information about places thought to have spiritual power, places where they collect important plants, animals, or minerals, or carry out religious activities, and places where their ancestors left things—like their bodies. But cemeteries aren't the only such sensitive places that they may want to protect from disturbance. To the extent allowed by whatever freedom of information laws apply to our situation, we need to try to respect these wishes. We may want to keep information confidential ourselves to avoid giving people "treasure maps" on which archaeological sites represent "dig here" signs. Figuring out what information needs to be kept confidential, and how we're going to do it, is an important part of scoping a survey.

Background Research

Background research is usually done as part of scoping and designing field surveys, but it also is a discovery method in itself. It usually precedes field survey, but it also often continues during the fieldwork, in a dynamic relationship with it.

Background documentary study as a prelude to field survey may yield information on the specific locations of particular archaeological sites, but this is not its most important purpose. The major function of background research is to allow the development of expectations about:

- What kinds of sites may be expected in the study area?

- What environmental, social, and historical factors may have influenced their distribution, and hence in what sorts of locations can sites be expected?

- What will they look like if and when they are found?

- What cultural processes and patterns do they reflect, and what is their possible significance for research?

- What other social or cultural values may be attributed to them above and beyond their research value?

- What special kinds of expertise or special methods may be required to locate, identify and evaluate them?

Background research requires the examination of many kinds of data sources, for example:

- People in the unrecorded (prehistoric) past may have responded to environments that no longer exist (e.g., the pluvial lake in Griffin Valley). As a result, knowledge of the area's geography, geology, and possible paleoenvironments is important in predicting where sites may occur.

- Historic period settlements may have been located or abandoned in response to particular documented events (e.g., the Indian "uprising" in Griffin Valley), along particular transportation corridors, or as a result of technological developments (e.g., the automobile) or large scale patterns of social change (e.g., the immigration of eastern Europeans to America). Data on historic patterns of land use, economic change, social interaction and technological innovation are therefore important in predicting where historic sites will occur, what they will look like, and what their associations will be with the broad patterns of regional and national history.

The most general kind of background research needed for a survey project is becoming grounded in pertinent anthropological, sociological, or geographic history and practice. Effective surveyors have sufficient background to evaluate archaeological sites in light of the data they contain about the human past. A comprehensive survey requires breadth of knowledge; the surveyor must be at least generally familiar with all types of research problems that might be addressed using the types of sites that may occur in the area.

Background research can also inform us about what kinds of non-research significance may be attached to archaeological sites and other places—such as structures, landscapes, landforms, plants, animals, water, and so on. While it may be someone else's job to document these places and their significance, an archaeologist needs to know about them because the values that people ascribe to a place may affect how

they react to the way an archaeologist approaches, documents, or expresses himself about it. For example, if an Indian tribe expresses respect for the dead by never mentioning them, and by trying never to disturb them, its members may react badly to an archaeologist's painstaking report on a human skeleton found in a rodent burrow during a survey.

Finding Sources

Where do we go for background data? Experienced archaeological surveyors will develop a variety of sources in the course of their work, most of which may be peculiar to the area under study. Here are some examples of places to look.

The National Park Service maintains a central repository of *National Register of Historic Places- Listed Properties and Determined Eligible Properties* available on-line at http://www.cr.nps.gov/nr/research/ index.htm. Offices of historic preservation or state archaeologist offices established in most states and territories under provisions of the National Historic Preservation Act are charged with responsibility for securing and maintaining statewide inventories of archaeological sites, historic places, historic points of interest, historical landmarks, and historic property data files. In some states, non-profit historical societies or archaeological societies have played lead roles in securing or implementing the federal grants under the act, and thus play roles in producing and maintaining clearinghouse repositories. Some federal agencies keep parallel and entirely independent records of archaeological sites and surveys, and so you may need to visit (among others) national park offices, national forest supervisor or district offices, or Bureau of Land Management resource area offices in your region.

For data on present and previous environments, in addition to published material in the geological, geographic, and ecological literature, there may be unpublished studies in local university and college geography, geology, or biology departments and in county and regional planning offices. The Bureau of Land Management and the Forest Service conduct studies and compile data on past and present conditions of land under their jurisdiction. County and state offices of the Department of Agriculture's Natural Resource Conservation Service has detailed studies of soils from which previous environments can be reconstructed. Commercially available aerial and satellite imagery can be used to identify the distribution of present plant communities and, in some cases, to detect evidence of previous environmental conditions. Much of this material is available on the internet, either for free or at relatively low cost.

For data on local history, published and worldwide web-based county and town histories and historical atlases provide good starting points but are seldom adequate in themselves. Academic social and economic histories may be important for establishing general patterns of social change. Most communities have historical societies or museums that maintain old maps, diaries, journals, newspapers, and similar sources of primary data, many of which are being digitized and made available on the web. Because utilizing such sources can be a massive undertaking, it's important to have some definite plan in mind before we start, as a basis for prioritizing. We don't need to be concerned with everything that happened in the study area, but only with those things that might be directly or indirectly reflected in or on the ground. On the other hand, even the simplest anecdotal accounts of life in an area may reveal things about social and economic conditions and may contain clues to locations where specific activities or events occurred. There is no escaping the fact that for any area with a history, a complex pattern of environmental change, or a considerable body of pertinent historic, archaeological, or anthropological literature, background documentary research is likely to be a complex and lengthy operation. A common weakness of archaeological survey projects is the archaeologist's failure to budget enough time and funds for such work.

Historic maps are often housed in state libraries or county or regional historical societies or museums. Historic Government Land Office (GLO) plat maps or PLSS records are available for most areas of the midwest and western states and these, along with historic homestead land patents, are managed and housed regionally by offices of the Bureau of Land Management.

Not all social groups have equal representation in the written record. Published histories, particularly those published before the late 20th century, tended to emphasize the activities of society's dominant segments, and the generally higher literacy rate among members of the upper class means that they tended to be better represented in the documentary record as well. Often the only sources of data on less

dominant social groups are oral—what the descendants of such groups can tell us. A systematic program to interview such people may be necessary to gain a full understanding of the area's social history and to identify possible sites of importance to the various segments of the community.

Community input and consultation is also an important and growing concern in the conduct of surveys. In the course of historical research and interviews, it is important to be alert to evidence that sites—archaeological or otherwise—in the area may have non-archaeological significance to the people being interviewed or others. In Griffin Valley, for example, we might expect that the descendants of the local Indian tribe would feel strongly about the sites of their last battle, the mass grave, and the ridgetop ritual area. Their feelings would define the significance of the sites and indicate some of the things the tribe should be consulted about during the survey. The background research might also indicate that some properties in the area have architectural value, are associated with broad historical trends or specific events, or contribute to the ambience or character of a community or landscape. This might in turn suggest that the survey team include not only archaeologists but architectural historians, historians, geographers, anthropologists, or sociologists.

A basic understanding of the available ethnographic and archaeological literature on the area is vital to the success of the survey. We don't necessarily need to know everything there is to know about local cultural groups and their archaeological leavings, but we do need to know what, if anything, can be said about the settlement patterns, patterns of social interaction, economic practices, and archaeological site types associated with different periods in the area's history and prehistory.

Remote Sensing

One more prelude to putting boots on the ground in an archaeological survey is often some kind of remote sensing. "Remote sensing" is the name given to the extensive set of technologies that permit us to identify things at a distance. The term actually applies both to methods used to identify things on or under the ground from an elevation, usually in the air or in space, and to methods used to identify things under the ground from the ground surface. It also applies to finding and characterizing things on or under the sea floor (or the bottom of any other water body) from the water surface or some point in the water column between the surface and the bottom. In deep water, remote sensing may be the only way to do archaeological survey at all. For the moment, let's focus on the use of aerial and space-based remote sensing as a means of guiding, prioritizing, and planning field survey.

One important use of remote sensing is in the reconstruction of past environments. This is useful in its own right as a way of generating data to help us understand how and why people lived the ways they did in the past, and it can also be helpful in locating archaeological sites. If we can figure out where the buried river channel, old lakeshore, or closed-up spring was, we can search those locations for sites representing their previous human uses.

Remote sensing can also identify sites directly. In Griffin Valley, Dr. Beakey employed a simple form of remote sensing when he caught a glimpse of the field scars on the valley floor in the slanting rays of the setting sun. Aerial photography, under differing conditions of light and vegetation, after fields have been plowed, or after rains can similarly reveal phenomena invisible on the ground. More sophisticated types of sensing, such as aerial magnetometry, multispectral imaging, airborne television, thermal infrared scanning, and radar are also being employed routinely in archaeological survey.

Prefield Agreements and Actions to be Taken on Discovery of Human Remains

Background research should also extend to making yourself and your team fully cognizant of all pertinent standards, laws, and policies. Even though national and regional archaeological societies often fail to identify "knowledge of pertinent state and federal law and policy" as a specific code-of-conduct issue, working knowledge of guidelines, laws, and policies applicable to your planned activity is an ethical and professional responsibility of the highest order. Sources for federal and state professional archaeological standards and guidelines are listed in Appendix A. Be aware that—as an archaeologist—you have no special dispensation from laws pertaining to the discovery of human remains, and indeed you are generally expected to exercise a special responsibility to uphold federal and state provisions concerning the discovery and treatment of human remains. While many archaeologists are acutely aware of the Native American Graves Protection and Repatriation Act (NAGPRA) as it applies to existing collections, few also realize that NAGPRA contains strict protocols to be engaged after unanticipated field discovery of human remains on federal land. These protocols often differ markedly from those applicable if the find is on state or private land. Our Griffin Valley team learned these laws and where they might apply in our project area.

Our Griffin Valley team leader then contacted the Ebirt Tribe heritage liaison to discuss the tribe's preferences for actions to be taken in the event of unanticipated discovery of human remains during the survey. Based on this conversation we wrote up a draft *Prefield Agreement for the Treatment of Unanticipated Human Remains* and forwarded this to the heritage liaison at the tribal headquarters. The Liaison commented on the draft and we responded, amended accordingly, then visited the liaison at the tribal office to secure her signature on the agreement.

Artifact Collection and Curation Policy

Our Griffin Valley survey team will also enter the field with a clear and concise collection strategy which lists a set of key artifact types to be collected and curated by the investigation and gives a prospectus on the kinds of scientific studies to be undertaken. The collection strategy has been carefully linked to the research design by specifying how certain research questions are to be answered by collecting certain kinds of artifacts that can be subject to typological, metallurgical, dating, trace element, or biochemical studies that may reveal information about age, manufacture, trade patterns, and function. The Ebirt Tribe, landowners, and others with possible interests have been consulted about it, to make sure there are no major objections—for example, by the Ebirt, who as it turned out wanted any stone bowls found to be turned over to the tribe after they are described for the record.

We have also secured an agreement to curate and a letter of intent from the facility where the artifacts will be housed: the Indeterminate State University, Archaeological Curation Facility. This facility was a logical choice because it is the regional clearinghouse and already houses the field notes and collections recovered by Dr. Beakey and Loumington, so any future research is much more efficiently pursued by having all related collections in one place. Further, Indeterminate State University has an agreement for access endorsed by researchers representing the Indeterminate State Archaeological and Historical Society and a guiding principles agreement with local tribes which includes terms and conditions for accessing collections pertinent to their heritage interests. So, by housing our collection in this facility we are assured of keeping our findings accessible to all interested parties and allowing other, future investigators to check, amend, and augment our results.

Having secured this agreement, our team then turned to other interested parties. The collection policy was reviewed and approved by the lead federal agency, which sought to insure that immediate and long-term plans for collected artifacts were consistent with provisions of the code of federal regulations. The collection policy was also discussed and approved by landowners, who would otherwise retain ownership of the artifacts but agreed to sign a waiver relinquishing these rights and ceding ownership to the curation facility at Indeterminate State University. The collection policy was discussed with tribal and historical society partners and crew, who were consulted and kept fully informed about the intent to collect, the planned scientific studies, and the ultimate curation facility and its access protocols.

Limitations of Background Research

Background documentary research is an essential part of any survey program, but unless it reveals that the area has been subjected to highly intensive archaeological survey, or that it's virtually impossible for archaeological sites to exist there, it can't eliminate the need for some type of inspection in the field. A documentary record that is representative of all social groups, activities, and time periods in the history of an area would be a great rarity. But such a record would still require field verification because people don't always do what they say they will. And much of the documentary record of any area's history is based on people's memories, which are incomplete and subject to error and unintentional bias. The archaeological record in and on the ground is incomplete too, biased in its own ways and certainly subject to mis-representation, but it still represents an independent source of data on an area's history. A variety of methods can be used to seek out and begin to document that record—which, of course, is what archaeological survey is all about.

But checking all the background data and consulting everyone who can be consulted still doesn't tell us specifically what archaeological sites lie in Griffin Valley. To learn that, fully informed by our background research, we need to put boots on the ground.

Types of Archaeological Field Survey

Chapter 10

Having concluded that we want to do a field survey of Griffin Valley, we now have to decide how to do it. There are several common approaches to survey field work, which yield different kinds of results.

We'll begin by distinguishing between what we call *exclusive* and *non-exclusive* survey. Exclusive survey focuses on some areas, zones, or contexts and excludes all others. *Uncontrolled–exclusive survey* does this without specific controls, responding to intuitive or some non-systematically applied learned cues. *Controlled–exclusive survey* focuses on certain areas and excludes others in a systematic manner with biases, goals, and strategies clearly spelled out and positively linked to areas selected for survey and others selected for no survey. In a *non–exclusive survey*, no portion of the study area is excluded from inspection; survey coverage is "complete." Coverage may be complete at a number of different levels of intensity, however, and the level of intensity will naturally affect the probability of identifying all archaeological sites.

In this chapter, we'll examine special kinds of survey involving distinctive packages of methods and methodology that are more and more common dimensions of archaeological survey. Finally, we'll look at predictive survey for comprehensive planning

TYPES OF CONVENTIONAL ARCHAEOLOGICAL SURVEY

The Uncontrolled–Exclusive Survey

In an uncontrolled-exclusive survey, certain areas are *excluded* from inspection because one believes that they won't contain archaeological sites, and the decision to exclude such areas is made on the basis of *uncontrolled*—that is, unverified—assumptions. One of the most common uncontrolled assumptions used to structure archaeological surveys is that "people always

live near water." This is an intuitively attractive proposition, and it has led innumerable archaeologists to walk carefully up and down stream-banks while excluding from consideration open plains, hilltops, mountain sides and other areas. There are three general problems with this assumption. First, it is not true; second, it is vague; and third, even if it were true it would not be directly translatable into the statement "archaeological sites are always near water."

People do not always live near water. If defense is a major consideration, it may be worth the trouble to carry and store water at an easily defended position rather than expose oneself near a lake or stream.

Furthermore, what does "near" mean? Were the Clovis hunters in Griffin Valley camping "near" water when they carefully located their camp within sight and walking distance of the lake but not directly on it, in order to avoid discouraging game? The question: "how near is near" is clearly important if one is to use the "near water" assumption as the basis for structuring a survey.

The locations of water sources change: witness Griffin Valley, where the lake has dried up and the spring has closed. People used to live close to these water sources, and they certainly left things there that have come to be archaeological sites, but the sites are no longer near water.

Most important of all, perhaps, is the fact that human beings do more than just reside in an area, so they also produce sites that are not "living" sites and hence are not necessarily oriented toward water sources even if they *do* always live near water. In Griffin Valley people engaged in initiation rituals on top of mountains far from water; they also fought a battle, been buried, and searched for gold in such places.

The same observation applies to other commonly held assumptions like "people always live on or near good agricultural soils" and "people never live on slopes of greater than X degrees." Even were such assumptions one hundred percent correct, they would not be good predictors of location for the whole range of archaeological sites produced by a human population.

An uncontrolled-exclusive survey of Griffin Valley guided by the "near water" assumption would probably result in the identification of the first agricultural village, the historic refugee village, and the early seed-gathering camp; it might also result in the identification of the settler's cabin site if the surveyor were able to recognize and appreciate historic material. It would probably miss everything else. It might not be a bad thing to do as a sort of casual reconnaissance while planning a more extensive survey—or just for the hell of it—but we shouldn't portray it as a thorough inspection of the valley.

The Controlled–Exclusive Survey

At the opposite end of the spectrum of survey efficiency is the controlled-exclusive survey. In such a survey, one has sufficient information on an area to make solid and defensible judgments about where archaeological sites may and may not be. Taking Griffin Valley as an example:

- If we knew there had been a pluvial lake, and could identify its shoreline elevation, and knew about hunter-gatherer hunting strategies, we would carefully search a band of territory around the shoreline, extending back into sheltered areas where camps might be located, and would probably find both the mammoth kill site and the Clovis camp.

- If we knew the geological history of the area, and understood how to locate extinct springs and water sources, we might examine each such location and find the early seed grinding village as well as the sites along the extant waterways.

- If we knew that the area had experienced a period of population pressure and warfare during a particular prehistoric period, and had data on the locations of other defensive sites of the period in

other similar areas, we would probably explore the ridge tops and find the palisaded village.

- If we knew about ethnographic accounts of initiation rituals, or were told about such rituals by people of the Ebirt Tribe, or had data on the distribution of rock-art sites from systematic sample surveys, we would realize the need to check the rocky ridgetops, looking under overhangs.

- If we had carefully studied the history of the area, we should at least have ideas about where the first settler's cabin, the massacre, and the 1872 gold rush took place.

In each case, by knowing where sites are likely to be, we are able to direct our efforts to areas of high probability at the expense of those where such sites are not likely to occur. In order to do this with a reasonable assurance that we are not missing anything important, however, we must know our area very well and be able to demonstrate that our assumptions are correct. This requires a solid grasp of background data on local history, prehistory, and the natural environment. It also requires a good grounding in pertinent aspects of anthropological and geographic theory, from we can make educated guesses about how people in the past behaved and where different behaviors would most likely be localized. We also need firsthand data on what has been found in areas similar to the one we are inspecting, and we need to look at a representative sample of our area's landscape.

It is sometimes assumed that after working for a long time in an area, an archaeologist has automatically gained enough understanding of the area's archaeology to undertake controlled-exclusive surveys. This assumption is risky in the absence of some way to test and verify the archaeologist's beliefs about where sites will be found. Such beliefs, when not rigorously tested, can become self-fulfilling prophesies. To return to the "near water" assumption for a moment: if one always looks for sites along streams, and never looks elsewhere, one is obviously going to find sites only along streams. However many years one spends finding sites along streams and never looking for them anywhere else, one will never prove that sites occur only along streams.

Not all archaeologists are particularly interested in testing their assumptions, and there is nothing necessarily wrong with that. If one is interested in conducting research only in types of sites that always occur along streams, there is clearly no reason to look elsewhere; it does not matter that other sites, of types irrelevant to one's research, exist in other settings. For management purposes, however, one needs to adopt a broader viewpoint. One has to try to identify all types of sites that may be eligible for management—for example in the United States, by being eligible for the National Register. The fact that the longest-resident or most eminent archaeologist in the area is not interested in some classes of sites does not necessarily mean that they are universally unimportant, and hence do not need to be managed.

We do not mean by the above to denigrate the importance of experience. An archaeologist who is experienced in an area should be much better at locating and interpreting archaeological sites there than someone new to the region—provided the experienced archaeologist has gained some humility in the course of her experience, and doesn't assume that she knows more than she really does. And someone who is not trained in archaeology but who knows the land well—the farmer who has plowed it scores of times, the cowpuncher who has viewed its contours in every season and in kind of sunlight and shadow, the artifact collector who has scoured it for relics—may know things about the area's archaeological sites that no one can find out in a short-term study, however rigorous it may be. Experience needs to be put to work in figuring out where different kinds of sites are likely to be found, but the results of that figuring out should always be tested somehow. That testing is what makes a controlled-exclusive survey "controlled," what makes it more or less safe to exclude some areas from inspection.

In the case of Griffin Valley, of course, we don't have data that is sufficiently comprehensive and reliable to justify undertaking a controlled-exclusive survey.

Non-Exclusive Survey

In a non-exclusive survey, no portion of the study area is excluded from inspection; survey coverage is "complete." Coverage may be complete at a number of different levels of intensity, however, and the level of intensity will naturally affect the probability of identifying all archaeological sites.

The most obvious distinction among non-exclusive survey types is that between *non-exclusive surface survey* and *non-exclusive survey with subsurface exploration*. In conducting a non-exclusive surface survey one simply inspects the surface of the ground wherever this surface is visible, with no substantial attempt to clear brush, turf, deadfall, leaves, or other material that may cover the surface and with no attempt to look beneath the surface beyond the inspection of rodent burrows, cut banks and other exposures that one comes upon by accident.

A non-exclusive survey with subsurface exploration involves some definite effort to expose obscured surface conditions and/or to monitor subsurface conditions in a planned fashion. Various methods for subsurface exploration are discussed below and in Chapter 11.

Subsurface exploration may or may not be necessary in any given area, or in particular portions of an area. In planning a survey, we need to give thought to what sorts of subsurface exploration may be necessary. And we need to report what kinds of subsurface exploration we did and where we did it. When evaluating the results of a completed survey, a reviewer needs to be able to identify the extent and nature of subsurface exploration and to consider whether a failure to probe beneath the surface, or to have an adequate distribution of subsurface tests, may have resulted in a major failure to identify archaeological sites.

Another major distinction is between *non-exclusive survey with background research* and *non-exclusive survey without background research*. Background study of environmental data, historical sources and ethnographies will generally result in giving special attention to particular portions of the study area where special types of sites are expected to occur, and may result in the employment of special detection techniques in such portions of the area. For instance, in the case of Griffin Valley, as noted above, if our background research had informed us about the location of the old lakeshore, we would certainly give its contours close attention. We might conduct close-order shovel testing or some other means of being sure we hadn't missed anything in this high sensitivity area.

Types of Survey

Some Definitions:

Coverage = the location, description, and size of areas selected for survey.

Intensity = the methods and extent of effort used to search in the coverage area.

Systematic = methods applied according to a uniform plan and conducted uniformly in a specific coverage area.

Unsystematic = done without a specific plan. Methods at varied level of intensity. Coverage random, opportunistic, or intuitive.

Non-Exclusive Survey
Complete coverage. No portion of the project area is excluded and all areas receive the same level of intensity of survey.

Exclusive Survey
Incomplete coverage. Portions of the project area excluded. Focuses survey on certain areas and excludes other areas.

Types of Exclusive Survey

Uncontrolled–Exclusive
Unsystematic. No specific controls on coverage or intensity.

Controlled–Exclusive
Systematic focus on specific landforms and contexts, such as streamside terraces or flat-topped ridges. Reduced coverage on other landforms and contexts.

If we were able to do little or no background research—an unhappy situation but one that does sometimes occur—or if the area was so unknown that there was no background research to do, then we would employ uniform inspection techniques throughout the study area, insofar as we could. This would constitute a *non-exclusive survey without background research.*

A third distinction is between *non-exclusive deployed survey* and *non-exclusive gang survey.* In the former type, field crew members are deployed over the landscape in accordance with some kind of plan to ensure essentially total inspection of the land surface. In a survey with subsurface exploration, subsurface tests are also deployed at some sort of regular intervals. In a gang survey, the field crew moves through the area as a group or gang, spreading out informally in some places, bunching up in others, splitting and segmenting to check spots on either side.

Finally, we can distinguish between *non-exclusive comprehensive survey* and *non-exclusive special-purpose survey.* The former obviously means that one surveys in order to find all the types of archaeological sites present in the study area; the latter means that one surveys in order to identify some particular class of sites.

It is easy to say that for planning purposes in the absence of sufficient data to justify a controlled exclusive survey, one should always conduct a *non-exclusive comprehensive deployed survey with background research and subsurface testing.* As a general rule this is true. In particular circumstances, however, each of the types of survey described above may be appropriate or necessary. In many areas, such as the arid to semiarid western United States, the land surface is sufficiently well exposed in many areas, and soil formation is sufficiently slow, to permit the assumption that all reasonably identifiable archaeological sites can be found through surface inspection alone. Some areas may be so completely lacking in relevant documentary data that attempts at background research become exercises in futility. Under some environmental conditions (e.g., in narrow canyons or very thick brush) it may be impossible, overly hazardous, or simply unnecessary to deploy one's crew. If all the prehistoric sites in an area have already been identified, a comprehensive survey would clearly be wasteful and a special-purpose historic-site survey would be appropriate.

Thus, the exact type of survey undertaken will vary with the nature of the study area, and the techniques to be employed may vary substantially from place to place within the study area. Non-exclusive comprehensive deployed survey with background research and subsurface exploration is an ideal; while it is perfectly expectable that this ideal sometimes can't and need not be attained, it is important for the surveyor, and the surveyor's sponsor or client, to understand and fully report deviations from the ideal.

SPECIAL TYPES OF SURVEY

Thus far we have discussed a relatively traditional, straightforward type of archaeological survey: the search for sites in a fairly large, open area. Some variants on this theme require consideration.

Small-Area Surveys

The elaborate, multistage methods appropriate to a large area may be excessive for the survey of small areas such as small housing tracts, sewage treatment plant sites, and stock pond sites, but the general principles upon which they are based still apply. One needs to understand and control one's biases, do enough background research to know what to expect in the survey area, inspect the area systematically, and report carefully on what one does. It is usually not cost or results effective to develop complex research designs for small-area surveys or to engage in extensive background research. Ideally these surveys should be done with reference to a larger region of which the small survey area is part. In such instances the existence of organized archaeological groups in a region can be of great importance. If the archaeologists in a region can agree on a common set of research problems and procedures, can compile and share background data and establish uniform methods for conducting field survey (assuming that these are consistent with the state historic preservation plan and federal regulations) the conduct of small area surveys should be relatively simple and orderly. At this point statewide or other planning can play an important role. If the state, or a tribe, or an agency has reason to believe that many small-area surveys will be required in a given region, it may be efficient to sponsor the development of research designs and compilations of background data pertinent to the region and to assist the region's archaeologists in the development of procedures tailored to the area's specific characteristics.

Geoarchaeological Survey

Geoarchaeology—the application of geomorphic methods to archaeological problems—is important because the density and distribution of archaeological phenomena in many landscapes are only partly revealed by conventional archaeological survey methods, which concentrate on surface manifestations. In many areas, archaeological sites are deeply buried, and geoarchaeological methods are necessary to find these sites and to study the formation, erosion, and deposition of soils and sediments containing the archaeological deposits in order to improve our ability to predict buried site location and understand the processes leading to their burial (Goldberg and MacPhail 2006; Pollard 1999; Rapp and Hill 1998; Waters 2004).

Rivers and streams deposit and erode masses of sediment and create and remove habitable landforms and their associated habitation traces. Because of this, the surface archaeology in alluvial basins can be heavily shaped by geomorphic processes that occurred contemporary with human use. Older sites or specific site types located in landforms prone to flooding and erosion were buried or washed away and often under-represented in surface surveys. Not surprisingly, in many parts of the country there is an ongoing and vigorous debate about the effects of geomorphology on our understanding of past population (e.g., Artz 1996; Sheehan 1995, 1996). In many regions—and especially true of alluvial basins—rigorous application of geoarchaeological survey methods is the best hope we have for the discovery and evaluation of ancient archaeological phenomena (e.g., Holliday 1997).

In the context of an archaeological survey, geo-archaeology may be accomplished by a combination of laboratory and field studies. The former might include remote sensing to identify and analyze soils, evaluation of existing soil and soil-vegetation mapping or well and geological drill logs, and previous research on local geology. Preliminary field studies might be composed of a simple examination of soil profiles available in active stream cutbanks. However, the most productive use of geoarchaeology on archaeological survey involves the production of a prefield predictive model of expectation for the locations of buried archaeological phenomena followed by controlled, subsurface exploration of sensitive landforms.

Surveys in Urban Areas

Surveys in urban areas obviously present problems not encountered in Griffin Valley. Sites are buried not only under the ground but under pavement and buildings as well. While it is impossible to make judgments about subsurface conditions from surface indications, it is difficult, disruptive, costly, and often impossible to undertake subsurface testing.

It is a serious mistake—which has often proved costly—to assume that the mere fact of urbanization means that no archaeological sites can possibly survive. Commonly cities have developed in areas where prehistoric populations were also concentrated, and the development of a city itself leaves an archaeological record that is of great value to understanding the processes contributing to and affecting its growth (Orser 2004; Mayne and Murray 2004). The survival of archaeological sites in an urban environment depends on the construction history of the city itself. If extensive filling has taken place, or if buildings have been constructed on shallow foundations, preservation of subsurface remains may be quite good. If the history of city development has involved a great deal of deep-basement construction in areas that have not been deeply covered with fill, subsurface deposits may be completely disrupted. But even where a great deal of subsurface disturbance has taken place, important remnants of the past may still exist, deeper than the disturbance or in undisturbed pockets. Sometimes such remnants include parts of cemeteries or substantial buildings, which can be very costly to deal with if encountered during construction of a new project, so it's important to try to find them during the survey phase of planning.

Background research is of crucial importance in an urban survey. If there is some basis for predicting the distribution of prehistoric sites relative to their natural environment, the pre-urban environment of the city area can be reconstructed and one can then make reasonably educated guesses about where prehistoric sites will occur. While such predictions of site distributions can't be accepted without testing, in most cases predictions will allow such testing to be more focused than it would be otherwise, and hence more efficient and less costly.

Background research is particularly important for urban historical archaeology; a detailed study of old city maps, assessor's records, photographs and other illustrations, and written accounts should make it possible to plot the locations of previous buildings, streets, and areas of special activities. In dealing with industrial sites, knowledge of engineering principles and practices of the period under study is vital to the identification and interpretation of sites and features. Information on the social history of the community or neighborhood being surveyed is essential to the evaluation of its buildings and sites. In recognition of this fact, urban archaeologists are increasingly cooperating with social anthropologists, ethnohistorians, folklorists, and sociologists in studies that combine archaeology with oral history, documentary research, and ethnography (e.g., Alexandria Archaeology 1992; Mayne and Murray 2004).

Background research should also make it possible to sort out the developmental history of the city, distinguishing those areas that have been filled and/or built over only by light construction from those whose early subsurfaces have been subjected to extreme impacts.

With such information in hand, it should be possible to design a careful subsurface testing program that is concentrated on those locations where historic or prehistoric sites are most likely to have been and to have survived. At such a location there is almost always someplace to conduct subsurface testing. Because a mass of rubble is commonly encountered in such testing operations, and time is often short, backhoes and other mechanized tools are often used. Subsurface testing in an urban situation requires specialized skill and experience, both because the stratigraphy of the area tested is likely to be complicated and because of utility pipes, power lines, and toxic or hazardous material that may be encountered.

Even with substantial background research and adequate subsurface testing, the definition of archaeological sites in an urban survey is almost always less precise than is the case in rural areas. This element of chance is a part of any urban archaeological research and must be expected in surveys done for planning purposes. As a result of these uncertainties, construction project planners are often tempted to forego survey and testing and rely simply on monitoring construction. This can be a costly mistake, because significant archaeological sites found during construction—especially sensitive places like cemeteries—can bring major projects to a screeching halt and cost millions of dollars to deal with.

Surveys of Buildings and Structures

So far we have treated archaeological surveys as though they were exclusively concerned with sites lacking substantial structural remains. Obviously many areas surveyed do contain prominent structures, such as pueblos in the Southwestern U.S., and many recent standing structures can yield data that are important to archaeologists. A building in essence is a complex artifact, created and modified by people for economic, social, and cultural purposes. It is shaped by these purposes and reflects them. The original organization of a building reflects the builder's or architect's perceptions of how space should be organized for specific purposes that were important at the time of construction. Changes in its organization through time may reveal how purposes and perceptions have changed. Material left in a building, like the material left by the occupants of any archaeological site, can reflect the activities and concerns of the building's occupants or users. All can be fruitful subjects for archaeological study.

It is possible to get carried away with the archaeological value of buildings. Reduced to absurdity, one could argue that every building is an archaeological property, and that a survey should be conducted every time anyone contemplated adding or stripping wallpaper. Such an argument would serve no useful purpose. Professional judgment must be exercised in deciding whether a given building contains sufficient information of sufficient value to make it worthy of detailed study. In Griffin Valley, for example, the Ford house is one of the finest examples of Victorian residential architecture in the state. As such it undoubtedly qualifies for inclusion in the National Register on its architectural merits. It is also eligible for the register because of its association with the Ford family, important in the area's agricultural history. But it has been kept very tidy over the years; all trash has been removed from the premises, no graffiti decorate the walls, paint and wallpaper have regularly been stripped and replaced. As a result, the house can probably yield little archaeological data; little has been left to record changing styles, uses of space, size and organization of the residential unit, and so on. Because its yard and outbuildings have been maintained in a similar fashion, they are also of little archaeological value. There are only three potentially valuable features of the farmstead from an archaeological standpoint. One is an abandoned well behind the barn. Household trash was discarded on and near the well between 1895 and 1927, at which time a municipal dump was established and the Fords began hauling their trash to it. The well contains a discretely stratified sequence of material representing the family's economic ties, its food, drink, and medicine consumption, and preferences in disposable merchandise during the first 32 years of the ranch's existence.

Excavation of the well could provide valuable insights into how people lived and into the dynamics of the ranch's growth during the period. If it is likely that useful research questions could be asked about such topics, the well is a legitimate archaeological site and worthy of recording.

A second feature, or series of features, is represented today by shallow depressions marked in early spring by lush growths of native grass of a uniform rectangular shape and size. These phenomena indicate the precise locations of simple one or two-hole outdoor privies erected for the families who successively occupied the old Ford House. Owing to the private nature of the privies and the daily function they served, objects such as medicine and liquor bottles, tobacco smoking paraphernalia, watches and other artifacts can be expected to be found among their contents where they have long remained undisturbed.

Finally, although the outbuildings have for the most part been maintained in such a way that little remains in them from the past, there is one small room in an addition to the barn that has been unused and locked up for many years. During the 1920s it was used as a sort of bunkhouse for itinerant workers who came down from Canada and moved from ranch to ranch cutting and baling hay. Some of them left graffiti on the walls; others stuffed up chinks in the wall planks using pages from magazines and books they were reading. Some of their belongings were lost and filtered through the floorboards, or got stashed in a small crawl space above the ceiling boards. Itinerant Canadian hay balers no longer work in the state of Indeterminate or anywhere else in the U.S.; theirs is effectively a lost subculture. The room in the barn provides a sort of snapshot of that culture, and makes it a possibly significant archaeological "site."

Terrestrial Remote Sensing

Use of remote sensing technology to detail environments and help identify archaeological sites is described in Chapter 9, and we refer to this technology again here because perhaps the most important and ultimately most fascinating—to archaeologists—aspect of remote sensing is its growing capacity to detect subsurface variation within sites. Archaeological survey has recently enjoyed the development and diversification of on-the-ground instrumentation that allows the detection of subsurface features without ground disturbance, including electrical resistivity/conductivity, ground penetrating radar, and magnetometers, the latter representing the most active area of research and development. Magnetometers/gradiometers measure minute local variation in magnetic fields caused by things like variation in soil density and constituents, baked clay, and metal objects. The simple magnetometers/gradiometers of 15 years ago have been replaced by a wide array of technologies, with vapor chamber (proton, potassium, or cesium) and fluxgate magnetometers in most common use. Most of these applications are also notable for easy links to software capable of converting the data to two- or three-dimensional plots and other forms of digital graphics.

These machines can generally be assembled in two arrays. The magnetometer array uses sensing units side-by-side and the gradiometer array uses sensing units stacked vertically. The machine is then wheeled, carried, or dragged along at a constant height above the ground surface. The magnetometer array is more successful monitoring lateral variation in shallow phenomena, while the gradiometer array is more successful in monitoring vertical variation and greater depths (Burger 1992; Larson and Ambos 1997).

PREDICTIVE SURVEY

What is Predictive Survey?

Predictive surveys can be viewed as an attempt to build a database for a large land area without conducting a 100% non-exclusive survey. In a predictive survey one physically inspects only a fraction of the actual area of concern, and from this inspection—in the context of good background research—extrapolates to the entire area. Based on a predictive survey of high reliability (i.e., one that has been subjected to a number of carefully planned and executed tests) it should be possible to conduct controlled-exclusive surveys instead of non-exclusive surveys in advance of construction or land-use projects, thereby saving a considerable amount of time and expense.

Soon after the establishment of cultural resource management programs in the federal government, agencies with broad land management responsibilities learned the advantages of using predictive surveys in their general planning activities. For example, the U.S. Forest Service (Smith 1977), the Bureau of Land Management (Weide 1974), U.S. Army Corps of Engineers (Schiffer and House 1975), and the National Park Service (Benchley 1976; Dincauze and Meyer 1977; Gagliano 1977; King and Hickman 1973) all sponsored early predictive studies. Archaeologists and managers alike found these studies useful. However, over the next decade predictive models were not widely implemented because they remained expensive, involving difficult and labor-intensive methods, often using hand-drawn and hand-calculated efforts. After the mid-1980s, and the rapid development of Geographic Information System (GIS) software, predictive survey became the province of computer science and is now closely linked to GIS applications in archaeology. In fact, predictive modeling using GIS and other computer applications is a distinctive specialty with its own large body of specialized literature. A close look at this literature, however, will reveal a series of primary textbooks that rapidly begin to fade out-of-date as new software replaces old (Aldenderfer and Maschner 1996; Allen et al. 1990; Andresen et al. 1992; Barcelo et al. 1999; Judge and Sebastian 1988; Kvamme 1992).

As noted in Chapter 1, archaeology is increasingly partitioned into specializations that require students to dedicate themselves to a focused course of study in order to achieve a professional level of performance equal to that found in private, government, and academic practice. GIS and predictive archaeological survey is definitely one of these, and we would encourage any reader interested in the methods and applications involved to study one of the current textbooks on the subject (e.g., Mehrer and Wescott 2006; Wescott and Brandon 2000). Our purpose here is to discuss very generally how predictive sample surveys are done and how they can interlock with programs of comprehensive planning particularly those programs of statewide survey and planning undertaken by State Historic Preservation Officers (SHPOs) or

large, land managing agencies. For example, the Minnesota Department of Transportation employs the *Minnesota Archaeological Predictive Model* which features a number of proposed project applications and evaluative tests (see their web site at http://www.mnmodel.dot.state.mn.us/).

Research Design in Predictive Survey

A good research design is vital to the success of a predictive survey. The design should specify the general types of properties that will be sought and the criteria by which they will be evaluated. In addition, the design should set forth a strategy or strategies for sampling the study area. Such a strategy should provide a database for predictions that are rigorously testable (Ambler 1984). Only when a set of predictions has been tested and shown to represent accurately the actual distribution of archaeological sites can it be used confidently. Each new survey should be designed to constitute a test of the predictions until a high level of confidence has been reached.

Background Research in Predictive Surveys

Background historical, archaeological, and environmental research is vital to research design formulation. We can think of few things more impractical and dim-witted than trying to do a "predictive" survey without good background research. We'd argue, in fact, that any survey that makes effective use of background data is "predictive," and—more to the point—sample field survey is a way to test predictions based on background research. In other words, background research provides a basis for the first stage of the predictive study; based on background data one attempts to predict what kinds of sites will occur in the study area and where they will be found.

Studies of historic land-use patterns should enable one to predict at least general relationships between human activities and aspects of the natural environment. For example, background research in the general vicinity of Griffin Valley might result in the development of a table of predictions similar to that shown below. The same kind of predictive tables could be developed for prehistoric site-environment relations, as shown on page 94. Stratification of the sampling universe (discussed below) could then take these predicted relationships into account and fieldwork would test their accuracy.

It is important to emphasize that predictions from background research should be based on accurate settlement pattern data and well-grounded hypotheses about human-environmental relationships insofar as possible. An alternative approach is to take data on the distribution of known archaeological sites relative to features of the environment and then project a similar distribution of sites relative to all similar environmental elements in the study area. There are great dangers in this approach, however. Because we typically have no idea how information on known archaeological sites in any area was gathered (remember Beakey) it is seldom possible to state how representative the distribution of such sites may be of the actual distribution of all sites. It is safe to assume that the answer to this question, in most cases, will be "not very." In many instances, known archaeological sites are found to be consistently associated with modern roads and highways; this does not reflect the habits of historic and prehistoric people nearly so much as it does the habits of archaeologists. Were we to predict the distribution of sites in Griffin Valley from Beakey/Loumington data our result would be a map like the one shown on page 95—not inaccurate in its plotting of "high probability areas" as far as it goes, but leaving many areas of actual high sensitivity designated as "low probability areas."

Regardless of the sources used, one should never solely rely on predictions based on background research without field verification. In other words, one should never assume that because unverified predictions from background data indicate that an area has low archaeological sensitivity, a construction project planned for the area won't need to be surveyed. There are necessarily some pragmatic exceptions to the rule, however, as much field verification as possible should be completed before predictions are used as planning tools.

Field Work in Predictive Surveys

On the ground, the same methods are used for predictive survey as for any other kind of survey (see Chapter 11) but only portions of the whole study area are actually inspected. It is vital that the portions selected be representative of the whole. Much of the literature on predictive survey deals with the problems of choosing a sample to inspect that truly represents the important aspects of diversity within the environment of a study area.

Choices of Sampling Scheme

The system used to select the portions of the study area to be inspected is called the sampling scheme, and several types of schemes have been or are being used. The simplest scheme is that referred to by Mueller

Predicted types and distribution of historic sites in the Griffin Valley area.

Period	Social Group	Site Type	Distribution
Initial contact (Early 19th century)	Indians	Village sites	Same as terminal Stoneland prehistoric pattern
	Polish immigrants	Cabins with fields	Near fresh water sources but away from Indian sites on flat ground
Intensified immigration (Mid-19th century)	Indians	Refugee villages	Near fresh water but remote and relatively hard to reach: canyons high benches, etc.
	Whites	Defensive sites	Any high, rocky areas
		Cabins	Clustered together in loose communities for defense, near reliable water sources, flat ground away from high, rocky areas
Euro-American consolidation (Early 20th century)	Land barons	Major farm complexes	Ford farm

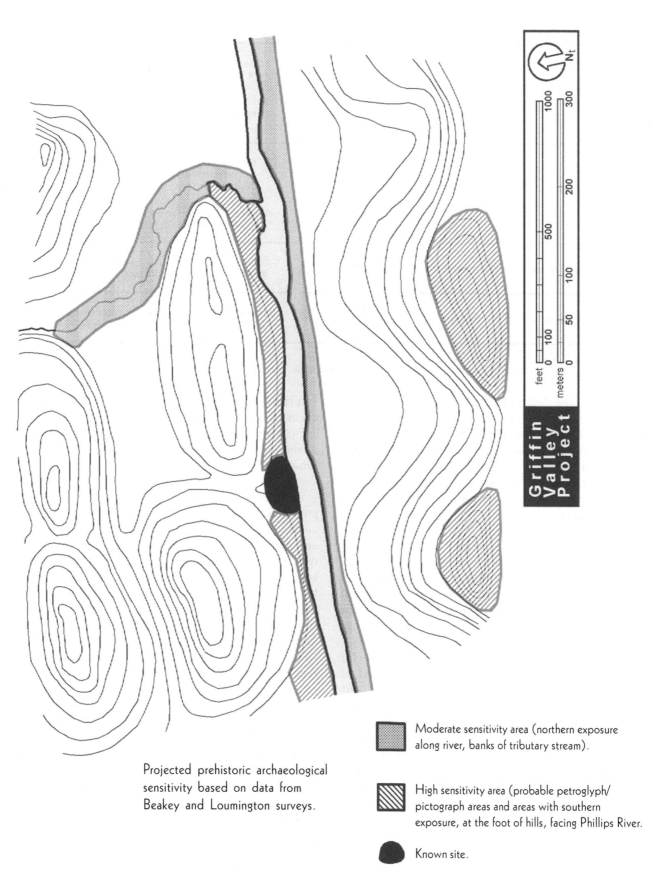

Projected prehistoric archaeological
sensitivity based on data from
Beakey and Loumington surveys.

Moderate sensitivity area (northern exposure
along river, banks of tributary stream).

High sensitivity area (probable petroglyph/
pictograph areas and areas with southern
exposure, at the foot of hills, facing Phillips River.

Known site.

Period	Social Group	Site Type	Distribution
Paleo-Indian	Clovis hunter	Campsites	Passes, hogbacks, near extinct springs, overlooking game trails, extinct lakes and streams, grazing areas
		Kill Sites	At the foot of cliffs, in box canyons, in extinct swamps, bogs, watercourses
Archaic	Hunter-gatherer	Semi-permanent	Near water sources and ecotones, especially chaparral/oak woodland ecotones
		Temporary seed-processing camps	In and immediately adjacent to sage communities, at bedrock exposures for grinding surfaces
Early Stoneland	Incipient agriculturalist	Semi-permanent to permanent villages	Generally same as Archaic; some oscillation away from ecotones and towards good agricultural soils
Late Stoneland	Proto-Irronole	Palisaded villages	Defensible areas within easy reach of fresh water and good agricultural land, but usually on high ground

Predicted types and distribution of prehistoric sites in the Griffin Valley area.

(1974:39) as the "grab sample," in which one simply "grabs" whatever data are available and makes predictions from them. We have discussed this approach above. It is not really sampling at all, and it is not recommended because of the uncertain representativeness of the sample and the possibility that field techniques used were not uniform among the various surveys.

"Simple random sampling" is a method of eliminating the biases from sample selection. The study area is first divided into equal-sized units (e.g., quarter-section "quadrats") which are assigned numbers. Units are then selected at random using a table of random numbers or some other objectifying device. Simple random sampling is useful in a homogeneous study area in which one lacks a basis for recognizing different sub-areas on environmental or other grounds. If the area contains different environmental zones that may have influenced settlement patterns, or if background research has suggested that particular sub-areas may contain particular types or densities of sites, simple random sampling essentially wastes this knowledge. Even if it is known that a given sub-area is likely to contain a particular type of site, one can't go and look unless the table of random numbers so directs.

"Systematic sampling" has similar weaknesses. In a systematic sample units are chosen at regular intervals on a grid so that one obtains a sort of checkerboard effect. Although this approach distributes sample units well over the study area, it does not guarantee that all sub-areas that may be of interest are sampled.

"Stratified random sampling" has become the most common general sampling scheme used in archaeology. In stratified sampling one first recognizes and delineates those sub-areas that are thought likely to contain different kinds or densities of sites or those attributes of the environment that are thought likely to have influenced settlement patterns. This, of course, is done on the basis of background research. Next, each sub-area, or "stratum," is divided into units, a sample of units is selected at random from each stratum, and the size of the sample is weighted to assure equivalent representation from all strata.

The figures laid out on pages 98–99 contrast the kinds of samples that might be selected from Griffin Valley using simple random, systematic, and stratified random sampling schemes, using environmental zones to define sampling strata.

Choices of Sampling Unit

Sampling units are the parcels of land chosen for inspection. Both unit size and unit shape are important because the units must be comparable and because their size and shape affect the number of units that can be selected and the efficiency with which they can be inspected. Archaeologists have traditionally used "quadrats" as their sampling units, selected from a grid superimposed on a map or air photo of the study area (pages 98–99 A–C). We can also survey "transects," which in many cases are more effective and less costly to inspect. A transect is a long, thin rectangle, perhaps one kilometer long and 100 meters wide. Transects are laid out in such a way as to cross-cut sampling strata, thus allowing the surveyor to observe greater variability than is possible with a quadrat; they are also easier to logistically coordinate, faster to cover on foot, and can be inspected with fewer people. Figure D on page 99 shows a stratified random transect survey selection at Griffin Valley.

Choices of Sampling Fraction

The sampling fraction is that portion of the total number of sampling units in the study area that is chosen for inspection. Sampling fractions used in predictive surveys have ranged from under one percent of the whole study area to over 50 percent (Mueller 1974:30). No reliable estimate can be made about the fraction needed to produce satisfactory predictions because there is no way of knowing at the outset the probability of sampling any one type of site in a given environmental context. Archaeologists have come up with different answers in different regions. However, our research design can make clear the nature of the data needed based on our prior understanding of the likely nature of the archaeological sites in the area and the management needs that prompt the study. Surface and subsurface inspections will also require different kinds of sampling methods and design and are likely to produce very different kinds of results, depending on the regional conditions influencing soil and sediment deposition and erosion.

As a rule of thumb, we suggest that for a large area like a state, a rough idea of site densities can probably be obtained from a sample of about one percent, combined with thorough background research. To obtain finer-scale predictions, define site types, or make predictions about smaller areas, much larger samples are needed. It is important to remember that the purpose of sampling is not the discovery of sites but the establishment of expectations about where sites will be and what they will be like. Sampling is not a

substitute for complete survey but one step in the survey process.

The Results of Predictive Survey

Predictive survey results are usually presented as maps portraying differential site densities or, in some cases, site-type densities. Predictions are made by noting the association of particular site types with particular environmental features within the sample units actually inspected, then projecting a similar distribution of sites to all equivalent environmental features within areas not yet inspected. A number of useful examples are available on the internet. Anderson (1995) reports a GIS-based effort to make a predictive model for archaeological site distributions in the Raccoon River greenbelt of west-central Iowa (see http://www.public.iastate.edu/~fridolph/archmod.html). Burtchard (2004) reports archaeological predictive modeling results for Mt Ranier National Park in Washington (http://www.nps.gov/archive/mora/ncrd/archaeology/contents.htm).

An association gains reliability if it was first predicted on the basis of background research and later verified by field survey. An association also gains credence if it is explicable once discovered (e.g., the camps of unrecorded gold miners are found near placer deposits) and if it is consistently reverified by further testing.

Hierarchical Predictive Surveys

For comprehensive planning, predictive survey may best be considered an ongoing process in which increasingly fine-tuned predictions can be made as more and better information becomes available. If the archaeologist continues to survey a new selection of sample units every year, he will eventually obtain a 100% sample. This is a rational goal for statewide comprehensive surveys and for federal agency surveys conducted under the National Historic Preservation Act. The advantage of predictive survey is that some useful data for purposes of planning in the entire study area become available almost immediately—for many parts of the country at least—and it is probable that all the information needed to carry out responsible preservation planning will be available before physical inspection has covered even 50 percent of the land.

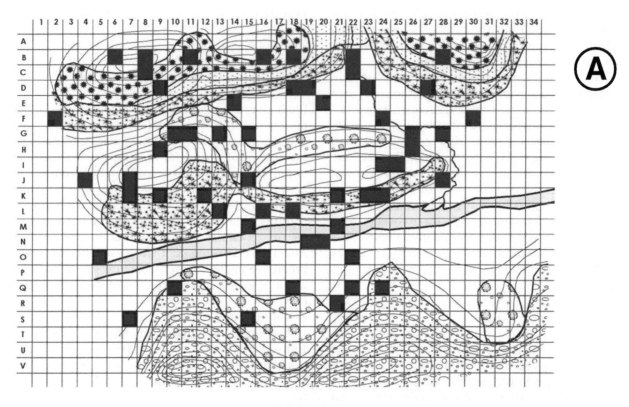

Alternative predictive sampling schemes, Griffin Valley area: (A) simple random, and (B) systematic random sampling.

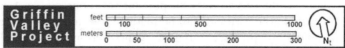

Griffin Valley Project

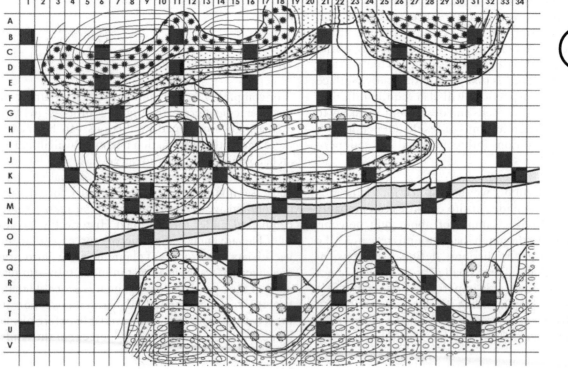

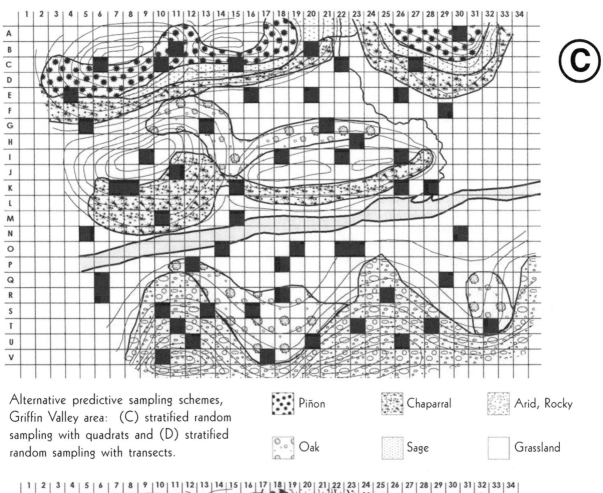

Alternative predictive sampling schemes, Griffin Valley area: (C) stratified random sampling with quadrats and (D) stratified random sampling with transects.

 Piñon   Chaparral   Arid, Rocky

Oak   Sage   Grassland

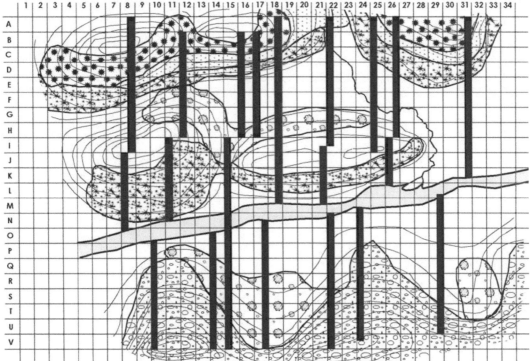

The ongoing predictive survey process is best conceived of as hierarchical, with refined predictions being built each year (or some other appropriate interval) through testing of older, less certain ones. The following basic phases can be projected:

Phase 1: Background research serves as the basis for developing preliminary predictive tables and maps. Field survey strata are established and a sampling scheme is developed.

Phase 2: Initial sample field work is undertaken. Depending on such factors as the size of the study area and the level of funding the sample fraction might be less than 1 percent or it might be as much as 10 percent; it will provide a rough check on the predictions developed from background research and result in a fact-based but still general prediction.

Phase 3: The sampling scheme is refined based on the results of Phase 2. The sample fraction can be increased with further field work and new background research may be undertaken to seek information on phenomena identified in the field. Results should include more refined predictive maps and data.

Archaeological Field Survey Methods

Chapter 11

In this chapter we'll discuss the essential operations involved in conducting an archaeological field survey, covering the basic methods that are common to most types of ground inspection.

BASIC METHODS

Establishing Pace Rate

During field survey there is a constant need for measuring space, especially calculating distance. Much of this is associated with basic logistics like establishing and maintaining transect width and measuring the distance between surface scrapes or shovel test pits. The creation of archaeological site sketch maps in the field also requires accurate estimates of the locations of features, objects, and boundaries (see Chapter 13). The method in most common use for rapidly calculating distance in the field is the *pace rate* (page 106).

The able bodied among us have a normal pace rate, and it tends to vary by the length of our legs, walking habits, and freedom of movement. Your normal pace is an easy gait that is both familiar and unconscious in most conditions, and it has a predictable length that can be measured and used to provide distance measurements.

To establish your pace rate, first lay out a 100-m or 300-ft reel tape in a straight line. You will have to stake the '0' end but on the opposite end most reel tapes come with a sharp mounting point on the reel and if you stick this in the ground and flip in the tape crank it will lock on one of the spokes, helping to keep the tape taut.

Once the tape is laid out, walk at a normal pace in a straight line for the length of the tape and count your paces. Stop at 100 m (for a metric pace rate) or 300 ft (for an English rule pace rate). To calculate your pace rate, divide your total paces by the distance, for example, 87 paces divided by 100 m equals 0.87, or 87 cm per pace. You can simply convert this to feet or inches to arrive at an English rule pace rate or vice-versa, depending on your circumstances. Use the pace chart provided here to calculate actual distance from your pace (pages 107–108). Double or triple as necessary to calculate distances greater than 100 m.

Try repeating the pace rate exercise on upslopes and downslopes to see how your pace rate changes. Note the upslope rate, downslope rate, and flat rate, and enter these on the inside page of your field notebook. Keep these differences in mind when you estimate distance in the field.

Transect Surveying

Most archaeological survey—and especially non-exclusive survey—is done using a basic transect approach. Transects consist of parallel lanes, with crew members each walking a single lane, side by side across the landscape and creating a coverage pattern that looks like the teeth of a comb (page 109).

Individual surveyors should concentrate on establishing and maintaining a path of movement within their own specific corridor. This is tough to do, especially because the method often calls for zigzagging back and forth within the width of your corridor to maximize coverage, so you change direction constantly and have to learn to "drift" toward the correct bearing. The best way to maintain transect integrity is for the crew chief to identify and communicate clearly to all crew members a specific bearing for transect alignment and establish and maintain the survey pace (page 110). Each crew member can then sight the announced bearing to align the transect, identify a distant object along the bearing, and then "drift" toward the object. Alternatively, if there is too much overgrowth and there are no distant lines of sight each crew member can "dial in" the announced bearing on his or her compass, and then monitor the compass regularly to maintain the correct track.

Remember, the crew chief will establish and maintain a walking speed. If you walk too fast and are way ahead, or too slow and are way behind, it is a strong indication that you are not practicing the same coverage intensity as your neighbors. Further, it will be harder for everyone to maintain transect integrity and thorough coverage. The crew person's job is to keep pace with neighbors.

Crew Chief Responsibilities

The crew chief's day is often much different and much busier than the crew member's. The crew chief's ability to "keep the parts in motion" during a survey is critical. The crew chief must: (1) identify and announce the transect bearing; (2) determine and coordinate coverage intensity; (3) establish, communicate, and maintain a coverage pace; (4) oversee delegation of responsibilities on discovery (see Chapter 13); (5) design and track survey organization; and, (5) maintain and review notebooks and records and oversee quality control.

Coverage Intensity

Coverage may be complete at a number of different levels of intensity, but the level of intensity will naturally affect the probability of identifying all archaeological sites. Different states and agencies have different definitions for coverage intensity, and you need to know these standards and either use them where they apply or be able to explain why you elected not to use them. Externally imposed standards aside, transect spacing depends on how much detail you think you need to record, balanced against practical factors of time and cost. Very closely spaced transects might place surveyors just 5 m (16.4 ft) apart, a sort of fine-toothed comb capable of teasing out lots of detail. Very loosely spaced transects might place surveyors 30 m (98.4 ft) apart. Tightly spaced transects give you increased coverage, but if the transects are too tight you risk duplicating effort and use lots of labor to cover a small area. Widely spaced transects cover lots of space with relatively little labor investment, but they

decrease coverage to the point where coverage may be inadequate for your original purpose.

Be familiar with pertinent guidelines, but think through this problem and come up with a method that is applied systematically and is appropriate to the context. For example, you might use tightly spaced transects on streamside terraces where you think human settlement and other activity is likely to have concentrated, intermediate spacing on ridgetops where background data suggest that transportation and other activities were staged, and widely spaced transects on steep slopes where sustained human activity is less likely to have occurred. Remember to qualify, quantify, systematize, and report the purpose and application of these coverage methods.

Remember to think these things all the way through to their logical conclusions and be clear about bias and sample control. If your investigation is focused on research questions involving trade and transport, one could justify, for example, more tightly spaced transects on the ridges where evidence of transport might be rather ephemeral and looser spacing on the stream banks where the big occupation sites are likely to smack you in the face.

Training Your Eye

It's common for a first-time archaeological surveyor to experience astonishment at an experienced surveyor's ability to find and identify hidden archaeological traces. This is no supernatural ability but a sign of solid training and acquired experience. This really speaks to the development of field skills that take time to acquire and are by nature regional. A storied California archaeologist, the late Adan E. Treganza, used to talk about the need to "calibrate your eyeballs" when you go to work in an unfamiliar area. Different shapes, sizes, textures, and materials populate the archaeological record, and even veteran surveyors can't see the obvious on their first couple of surveys, and then one day, bingo, they can.

Typically, veteran archaeologists have trained their eyes to "read" the landscape and "sweep" the ground. As for "reading" the landscape, a part of this ability relates to the veteran's familiarity with scholarly sources. Not just the crew chiefs and principals, but also the survey technicians should take the time to obtain and study maps of the survey area, and read anthropological and historical source materials. You can use ethnography and history to develop your expectations for site location and content. Remember that these expectations are potentially problematic because they are "particularistic"—in other words, dependent on the accuracy of the document and the behavior peculiar to the time period represented by the document—but they do provide an account of traditional land use in a past environment very likely quite different from the one we see today. These sources can also help you break through biases and assumptions you might adopt based on observation of current conditions.

Veteran archaeological surveyors also take the time to read sources on local historical and prehistoric archaeology. For the most part, the bulk of these sources and the ones that are most up-to-date are found in the "gray literature" of surveys done under contract for purposes of compliance with historic preservation laws. Studying these reports will help you become familiar with artifact types, including widespread general types and regional variants. This includes historical manufactured goods whose characteristics are often good markers of time, function, and social identity or status. However, don't take too many cues from this literature. Be aware that some of it is encumbered by shoddy methodology and by looking at the recent gray literature for guidance you risk replicating the status quo.

The other part of reading the landscape is the ability to see the potential land-use pattern of a given area, that is, the most efficient arrangement of a given adaptation to a given environment. Experienced surveyors will have the ability to spot where a homestead might best have been situated in a valley, or a hunting waylay positioned on a ridgetop. There are positive and negative sides to this ability. On the positive side, these expectations can be used to find more things and can be quantified in the form of a predictive survey. On the negative side, they also form biases and assumptions that can lead you to the expected and away from the unexpected. For example, train yourself to recognize things that are out of place in the natural environment and hence may indicate human activities: the walnut tree in the oak woodland; the rose bush next to the spring.

As for "sweeping" the ground (page 109), this is really a matter of training yourself to look down and training the eye to scan exposed ground. As you move along your transect, "sweep" with your eyes back and forth, side to side, covering an area at least 1 m (around 3 ft) on either side of your path. The eye should sweep at a fine scale, on the one hand, with eyes darting from object to object and lingering on things with a distinctive color, texture, or shape. On the other hand, the eye should also sweep at a broad scale, looking for larger properties or features in the vicinity including: distinctive landforms, like plow marks, mining ditches, or house pits; distinctive features, such as rock cairns or alignments, collapsed chimneys, or sleeping circles in the desert, and; distinctive associations, like unusual or nonnative trees, shrubs, or other plants that might signal past cultivation or plant management. The latter might include modified native trees bent or blazed to mark trails or trimmed to harvest bow wood.

Assaying Sensitive Landforms

Archaeological sites, features, and artifacts are not always easy to see, even for veteran field surveyors. The relative visibility of archaeological phenomena is often controlled by things we can't account for by surface survey alone, with vegetation—overgrowth, root mats, and duff overburden—and geomorphic factors—soil formation, deposition, and erosion—all diminishing archaeological visibility to different degrees in different contexts. And it is not just visibility but also preservation that might be affected by local soil chemistry, hydrology, and vegetation.

We think that all archaeological surveyors should have a working understanding of sediments and soils and landscape formation processes. Use this skill to learn where to look for buried archaeological sites and focus your efforts to assay subsurface deposits at these locations. At a minimum, never assume that what you see on the surface is the be-all and end-all. It is important to test and assay, especially on landforms that are likely to have traces of past human activity.

There are three methods in common use for assessing subsurface conditions on archaeological surveys. First, take advantage of all available exposures (page 112). Be especially attentive to creek banks, river banks, lake bluffs, or road cuts, which often provide unusually deep, long, and straight cuts through landforms where we are likely to find evidence of past human activity. These profiles can be inspected for archaeological traces but can also be studied to identify patterns in local landscape soil and sediment formation.

Once we understand these patterns we'll have a better sense of how deep sites might be buried on different kinds of landforms.

Second, use surface scrapes or shovel test pits to assay subsurface deposits (pages 113–114). In many areas, particularly in the arid to semiarid parts of the western United States, the land surface is sufficiently well exposed, or rocky, and soil formation is sufficiently slow, to permit the assumption that all reasonably identifiable archaeological sites can be found through surface inspection alone. Surface scrapes are still usually appropriate in these areas (page 114), particularly on landforms where there is poor to no surface visibility due to vegetation or duff. Surface scrapes are made using a trowel, hoe, or shovel to scrape back vegetation and overburden down to mineral soil. On landforms that your background research indicates are likely to bear traces of human activity, do scrapes at close intervals and in other areas scrape at broad intervals.

Shovel test pits (page 113) are small, square to round excavations generally measuring 40 to 50 cm (1.3 to 1.6 ft) across, with maximum depths established by regional convention. The excavated spoils are screened through 3 mm or 6 mm (^{1}/$_8$-in or ^{1}/$_4$-in) hardware cloth. STPs are excavated at regular intervals on survey transects to test the presence/absence of cultural materials and within site boundaries to mark the frequency distribution of artifacts of various sorts.

STP and surface-scrape intervals can be adjusted in order to meet research needs or regional guidelines applicable to your survey.

TRACKING AND ORGANIZATION

Maintaining a Field Notebook

Professionals generally maintain field notebooks to serve as a daily journal of scientific activities and observations. Hardcover or cardboard cover notebooks are appropriate, and we prefer quadrille rule or engineering rule papers that can be adapted equally to sketching or handwriting. If you are working in a tropical or littoral environment, or doing field work in the rainy season, then you will definitely need a water-resistant field notebook like those supplied by Rite-in-the-Rain.

As you prepare your field notebook, on the inside cover list your name, address, and other contact information, including a note indicating *"If found, please return to."* Define and describe the project on the first page. Start on a new page each field day. The header on each date should list in short form and systematic format: (a) the date; (b) the name of the investigation; (c) conditions (weather, lighting, wind); (d) kind of activity (site excavation, reconnaissance, mapping, etc.); (e) the crew chief; (f) the names of team members or co-workers; and, (g) the daily assignment.

Your first notes of the day should be done during the morning breakfast meeting or initial on-site orientation, during which the team leader should describe the status of the investigation, weekly and daily goals, and individual or team assignments. It should also be clear how your specific daily assignment fits into these goals. The bulk of your notes should pertain to daily, ongoing activities relating to these goals. Make sure that these notes are enriched by well-labeled sketches and sketch maps, including artifacts, features, or stratigraphic profiles. Use the afternoon or evening time reserved for note-taking to clean up your notes, redraw illustrations, check with co-workers, check site and transect records, and follow up with other observations and interpretations. Examples of survey field note formats are provided in Appendix D.

Daily and Transect Records

Each team should be responsible for completing a daily record including a coverage map, map coordinates and a description of the area covered, team member names, an acreage estimate, work hours, and a discussion of the conditions affecting survey. The daily record should also list the archaeological resources observed and the status of documentation at the end of the field day. Examples of daily and transect records are provided in Appendix C.

Calculating Survey Rates

As you work in different regions and under different conditions, keep notes on relative survey rates. How many acres did your team survey? The total acreage covered per day divided by crew size gives you an "acres per person-day" rate (acres/PPD). You'll find that the most important factors affecting acres/PPD rate are: (1) site density, (2) relative proportion of low versus high intensity survey, and (3) relative proportion of low versus high surface visibility.

Survey Strategies

One of the best ways to improve the efficiency of your survey is to anticipate bottlenecks and invent creative logistical solutions. Here are some examples:

> *Split the day*: Do half your work by midday. Sweep out in the morning, break for lunch, then sweep another set of transects back to your starting point in the afternoon.
>
> *Leapfrog reconnaissance*: When surveying alignments that have reliable vehicle access, such as roads or levees, split your crew into two teams using one vehicle with two sets of keys. Team 1 gets dropped off by Team 2. Team 2 drives ahead, for example one mile, then surveys ahead one mile. Team 2 should catch up and drive ahead two-miles, lagging Team 2 by one mile, and so on. Make sure you maintain some workable way to communicate, via cell phone or handset and have a plan for various kinds of emergencies.

It's also a good idea when surveying steep slopes to investigate upslope and downslope access and try to arrange vehicles and responsibilities to enable most of your crew to *finish downhill*. When it's hot, *start early and end early*.

Measure Your Pace

Use pacing to measure distance when you make a sketch map or to track spatial interval on transects, surface scrapes, or shovel test pits.

It's best to use a normal pace when you measure distance in the field. So, establish your pace rate using a normal pace.

For a Metric-Rule Pace Rate:

Lay out a 100-m tape. Start at 0 m and stroll at a normal pace to 100 m. Remember your total paces.

Divide the actual length paced by the number of paces to arrive at your pace rate. For example, if you went 125 paces, then:

100 m ÷ 125 paces = 0.80 m
a pace of 80.0 centimeters

Repeat the exercise on slopes and flats to arrive at an average.

For an English-Rule Pace Rate:

Convert the result to English rule (0.80 m = 2.62 ft) or repeat the exercise using a 300-ft tape.

Those with lanky legs may prefer to train themselves to pace 1.0 m, skipping the cipher.

Find your pace rate across the top and number of paces down the side to arrive at the actual distance. To speed things up, use a highlighter to mark your pace rate column.

Pace Chart, 0–50 m

Pace Rate (m)

Number of Paces	0.50	0.55	0.60	0.65	0.70	0.75	0.80	0.85	0.90	0.95	1.00
2	1.00	1.10	1.20	1.30	1.40	1.50	1.60	1.70	1.80	1.90	2.00
3	1.50	1.65	1.80	1.95	2.10	2.25	2.40	2.55	2.70	2.85	3.00
4	2.00	2.20	2.40	2.60	2.80	3.00	3.20	3.40	3.60	3.80	4.00
5	2.50	2.75	3.00	3.25	3.50	3.75	4.00	4.25	4.50	4.75	5.00
6	3.00	3.30	3.60	3.90	4.20	4.50	4.80	5.10	5.40	5.70	6.00
7	3.50	3.85	4.20	4.55	4.90	5.25	5.60	5.95	6.30	6.65	7.00
8	4.00	4.40	4.80	5.20	5.60	6.00	6.40	6.80	7.20	7.60	8.00
9	4.50	4.95	5.40	5.85	6.30	6.75	7.20	7.65	8.10	8.55	9.00
10	5.00	5.50	6.00	6.50	7.00	7.50	8.00	8.50	9.00	9.50	10.00
11	5.50	6.05	6.60	7.15	7.70	8.25	8.80	9.35	9.90	10.45	11.00
12	6.00	6.60	7.20	7.80	8.40	9.00	9.60	10.20	10.80	11.40	12.00
13	6.50	7.15	7.80	8.45	9.10	9.75	10.40	11.05	11.70	12.35	13.00
14	7.00	7.70	8.40	9.10	9.80	10.50	11.20	11.90	12.60	13.30	14.00
15	7.50	8.25	9.00	9.75	10.50	11.25	12.00	12.75	13.50	14.25	15.00
16	8.00	8.80	9.60	10.40	11.20	12.00	12.80	13.60	14.40	15.20	16.00
17	8.50	9.35	10.20	11.05	11.90	12.75	13.60	14.45	15.30	16.15	17.00
18	9.00	9.90	10.80	11.70	12.60	13.50	14.40	15.30	16.20	17.10	18.00
19	9.50	10.45	11.40	12.35	13.30	14.25	15.20	16.15	17.10	18.05	19.00
20	10.00	11.00	12.00	13.00	14.00	15.00	16.00	17.00	18.00	19.00	20.00
21	10.50	11.55	12.60	13.65	14.70	15.75	16.80	17.85	18.90	19.95	21.00
22	11.00	12.10	13.20	14.30	15.40	16.50	17.60	18.70	19.80	20.90	22.00
23	11.50	12.65	13.80	14.95	16.10	17.25	18.40	19.55	20.70	21.85	23.00
24	12.00	13.20	14.40	15.60	16.80	18.00	19.20	20.40	21.60	22.80	24.00
25	12.50	13.75	15.00	16.25	17.50	18.75	20.00	21.25	22.50	23.75	25.00
26	13.00	14.30	15.60	16.90	18.20	19.50	20.80	22.10	23.40	24.70	26.00
27	13.50	14.85	16.20	17.55	18.90	20.25	21.60	22.95	24.30	25.65	27.00
28	14.00	15.40	16.80	18.20	19.60	21.00	22.40	23.80	25.20	26.60	28.00
29	14.50	15.95	17.40	18.85	20.30	21.75	23.20	24.65	26.10	27.55	29.00
30	15.00	16.50	18.00	19.50	21.00	22.50	24.00	25.50	27.00	28.50	30.00
31	15.50	17.05	18.60	20.15	21.70	23.25	24.80	26.35	27.90	29.45	31.00
32	16.00	17.60	19.20	20.80	22.40	24.00	25.60	27.20	28.80	30.40	32.00
33	16.50	18.15	19.80	21.45	23.10	24.75	26.40	28.05	29.70	31.35	33.00
34	17.00	18.70	20.40	22.10	23.80	25.50	27.20	28.90	30.60	32.30	34.00
35	17.50	19.25	21.00	22.75	24.50	26.25	28.00	29.75	31.50	33.25	35.00
36	18.00	19.80	21.60	23.40	25.20	27.00	28.80	30.60	32.40	34.20	36.00
37	18.50	20.35	22.20	24.05	25.90	27.75	29.60	31.45	33.30	35.15	37.00
38	19.00	20.90	22.80	24.70	26.60	28.50	30.40	32.30	34.20	36.10	38.00
39	19.50	21.45	23.40	25.35	27.30	29.25	31.20	33.15	35.10	37.05	39.00
40	20.00	22.00	24.00	26.00	28.00	30.00	32.00	34.00	36.00	38.00	40.00
41	20.50	22.55	24.60	26.65	28.70	30.75	32.80	34.85	36.90	38.95	41.00
42	21.00	23.10	25.20	27.30	29.40	31.50	33.60	35.70	37.80	39.90	42.00
43	21.50	23.65	25.80	27.95	30.10	32.25	34.40	36.55	38.70	40.85	43.00
44	22.00	24.20	26.40	28.60	30.80	33.00	35.20	37.40	39.60	41.80	44.00
45	22.50	24.75	27.00	29.25	31.50	33.75	36.00	38.25	40.50	42.75	45.00
46	23.00	25.30	27.60	29.90	32.20	34.50	36.80	39.10	41.40	43.70	46.00
47	23.50	25.85	28.20	30.55	32.90	35.25	37.60	39.95	42.30	44.65	47.00
48	24.00	26.40	28.80	31.20	33.60	36.00	38.40	40.80	43.20	45.60	48.00
49	24.50	26.95	29.40	31.85	34.30	36.75	39.20	41.65	44.10	46.55	49.00
50	25.00	27.50	30.00	32.50	35.00	37.50	40.00	42.50	45.00	47.50	50.00

Pace Chart, 51–100 m

Find your pace rate across the top and number of paces down the side to arrive at the actual distance. To speed things up, use a highlighter to mark your pace rate column.

Pace Rate (m)

Number of Paces	0.50	0.55	0.60	0.65	0.70	0.75	0.80	0.85	0.90	0.95	1.00
51	25.50	28.05	30.60	33.15	35.70	38.25	40.80	43.35	45.90	48.45	51.00
52	26.00	28.60	31.20	33.80	36.40	39.00	41.60	44.20	46.80	49.40	52.00
53	26.50	29.15	31.80	34.45	37.10	39.75	42.40	45.05	47.70	50.35	53.00
54	27.00	29.70	32.40	35.10	37.80	40.50	43.20	45.90	48.60	51.30	54.00
55	27.50	30.25	33.00	35.75	38.50	41.25	44.00	46.75	49.50	52.25	55.00
56	28.00	30.80	33.60	36.40	39.20	42.00	44.80	47.60	50.40	53.20	56.00
57	28.50	31.35	34.20	37.05	39.90	42.75	45.60	48.45	51.30	54.15	57.00
58	29.00	31.90	34.80	37.70	40.60	43.50	46.40	49.30	52.20	55.10	58.00
59	29.50	32.45	35.40	38.35	41.30	44.25	47.20	50.15	53.10	56.05	59.00
60	30.00	33.00	36.00	39.00	42.00	45.00	48.00	51.00	54.00	57.00	60.00
61	30.50	33.55	36.60	39.65	42.70	45.75	48.80	51.85	54.90	57.95	61.00
62	31.00	34.10	37.20	40.30	43.40	46.50	49.60	52.70	55.80	58.90	62.00
63	31.50	34.65	37.80	40.95	44.10	47.25	50.40	53.55	56.70	59.85	63.00
64	32.00	35.20	38.40	41.60	44.80	48.00	51.20	54.40	57.60	60.80	64.00
65	32.50	35.75	39.00	42.25	45.50	48.75	52.00	55.25	58.50	61.75	65.00
66	33.00	36.30	39.60	42.90	46.20	49.50	52.80	56.10	59.40	62.70	66.00
67	33.50	36.85	40.20	43.55	46.90	50.25	53.60	56.95	60.30	63.65	67.00
68	34.00	37.40	40.80	44.20	47.60	51.00	54.40	57.80	61.20	64.60	68.00
69	34.50	37.95	41.40	44.85	48.30	51.75	55.20	58.65	62.10	65.55	69.00
70	35.00	38.50	42.00	45.50	49.00	52.50	56.00	59.50	63.00	66.50	70.00
71	35.50	39.05	42.60	46.15	49.70	53.25	56.80	60.35	63.90	67.45	71.00
72	36.00	39.60	43.20	46.80	50.40	54.00	57.60	61.20	64.80	68.40	72.00
73	36.50	40.15	43.80	47.45	51.10	54.75	58.40	62.05	65.70	69.35	73.00
74	37.00	40.70	44.40	48.10	51.80	55.50	59.20	62.90	66.60	70.30	74.00
75	37.50	41.25	45.00	48.75	52.50	56.25	60.00	63.75	67.50	71.25	75.00
76	38.00	41.80	45.60	49.40	53.20	57.00	60.80	64.60	68.40	72.20	76.00
77	38.50	42.35	46.20	50.05	53.90	57.75	61.60	65.45	69.30	73.15	77.00
78	39.00	42.90	46.80	50.70	54.60	58.50	62.40	66.30	70.20	74.10	78.00
79	39.50	43.45	47.40	51.35	55.30	59.25	63.20	67.15	71.10	75.05	79.00
80	40.00	44.00	48.00	52.00	56.00	60.00	64.00	68.00	72.00	76.00	80.00
81	40.50	44.55	48.60	52.65	56.70	60.75	64.80	68.85	72.90	76.95	81.00
82	41.00	45.10	49.20	53.30	57.40	61.50	65.60	69.70	73.80	77.90	82.00
83	41.50	45.65	49.80	53.95	58.10	62.25	66.40	70.55	74.70	78.85	83.00
84	42.00	46.20	50.40	54.60	58.80	63.00	67.20	71.40	75.60	79.80	84.00
85	42.50	46.75	51.00	55.25	59.50	63.75	68.00	72.25	76.50	80.75	85.00
86	43.00	47.30	51.60	55.90	60.20	64.50	68.80	73.10	77.40	81.70	86.00
87	43.50	47.85	52.20	56.55	60.90	65.25	69.60	73.95	78.30	82.65	87.00
88	44.00	48.40	52.80	57.20	61.60	66.00	70.40	74.80	79.20	83.60	88.00
89	44.50	48.95	53.40	57.85	62.30	66.75	71.20	75.65	80.10	84.55	89.00
90	45.00	49.50	54.00	58.50	63.00	67.50	72.00	76.50	81.00	85.50	90.00
91	45.50	50.05	54.60	59.15	63.70	68.25	72.80	77.35	81.90	86.45	91.00
92	46.00	50.60	55.20	59.80	64.40	69.00	73.60	78.20	82.80	87.40	92.00
93	46.50	51.15	55.80	60.45	65.10	69.75	74.40	79.05	83.70	88.35	93.00
94	47.00	51.70	56.40	61.10	65.80	70.50	75.20	79.90	84.60	89.30	94.00
95	47.50	52.25	57.00	61.75	66.50	71.25	76.00	80.75	85.50	90.25	95.00
96	48.00	52.80	57.60	62.40	67.20	72.00	76.80	81.60	86.40	91.20	96.00
97	48.50	53.35	58.20	63.05	67.90	72.75	77.60	82.45	87.30	92.15	97.00
98	49.00	53.90	58.80	63.70	68.60	73.50	78.40	83.30	88.20	93.10	98.00
99	49.50	54.45	59.40	64.35	69.30	74.25	79.20	84.15	89.10	94.05	99.00
100	50.00	55.00	60.00	65.00	70.00	75.00	80.00	85.00	90.00	95.00	100.0

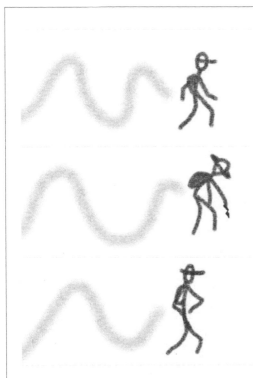

Transect Coverage

Transect Methods

Transects are parallel corridors, and each crew member walks one corridor at a time. The team walks side by side across the landscape making a coverage pattern that looks like the teeth of a comb.

Transect Coverage

Transect spacing on intensive coverage may vary from tight (5 m/16.4 ft) to wide (30 m /98.4 ft). Too tight and you'll duplicate effort, too wide and you'll risk good coverage.

Transect Tips

ï Concentrate on your own corridor.

ï Zigzag back and forth within the full width of your corridor to maximize coverage.

ï As you move along "sweep" back and forth, side to side.

ï The crew chief will establish and maintain a walking speed—the crew person's job is to keep pace with neighbors.

ï The crew chief will establish and communicate the transect bearing—keep your compass handy.

"*Sweep*" with your eyes at least 1 m (3.28 ft) on either side of your path of movement. Train your eye to track object to object and fix on objects with distinctive colors, shapes, or textures.

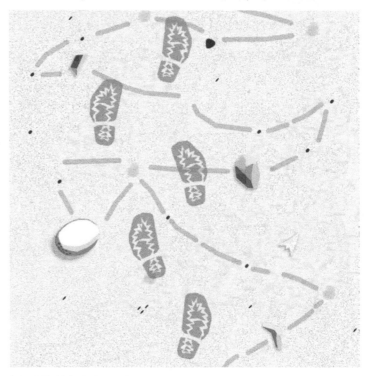

Crew Chief and Anchor

The crew chief sets and communicates a specific bearing for transect alignment and establishes and maintains the survey pace.

Maintaining Transect Integrity

Each crew member should sight the bearing to align her transect, identify a distant object along the bearing, and then walk toward the object.

Alternatively, each crew member can "dial in" the announced bearing on his compass, then monitor the compass to maintain the correct track.

On the trip back, turn 180° to the opposite bearing.

Anchor's Role

It is easy to inadvertently overlap parallel transects. The anchor—often the crew chief—can help avoid this problem. There are two methods:

In *method 1*, the anchor always rolls to the outside margin, flagging as she loses line-of-sight to the previous flag. Pinflag in barren land, or in wooded land use long strips of surveyor's flagging in bright colors, setting them high.

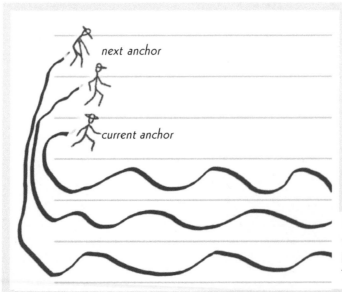

next anchor

current anchor

Turn and follow the flagging on the inside of the return transect.

In *method 2*, the anchor tracks and remembers the transect margin, taking the outside post heading out and the inside post heading in. The anchor's role rotates every cycle.

100%

Record Surface Visibility

Archaeological Surface Visibility

Surface visibility varies locally and seasonally. Forest floor duff is a persistent problem in conifer forests. Leaf litter obscures surface visibility in deciduous forests in the fall. Grass cover obscures the surface in the winter/spring in the far west and in the summer in the midwest and east.

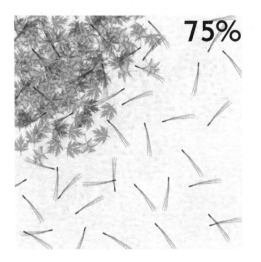

75%

50%

20%

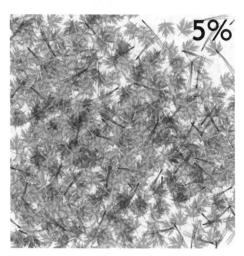

5%

Inspect Available Exposures

Examine Creek Banks

As you work your way through a survey area, make sure you cover available exposures along creek banks, river banks, road cuts, or lake shore bluffs.

These exposures are often ideal for archaeological purposes. Many provide deep profiles over a long cross-section through a landform where we might expect to see past human activity.

Look for Exposed Soils

Rodent mounds—the dirt spoils thrown out holes made by ground-dwelling rodents like prairie dogs and ground squirrels—are a great place to look for archaeological traces.

Root mounds from windfall trees are also very good places to look, and the roots torn from the ground often kick up large objects that rodents might tunnel around. Don't forget to look in the roots!

Shovel Test Pits

Shovel Test Methods

Shovel test pits (STPs) are small, square to round excavations generally measuring 40 to 50 cm (1.3 to 1.6 ft) across, with maximum depth depending on local geomorphology and the likely depth of cultural deposits.

STPs are excavated at regular intervals on survey transects to test for soil differences or test for the presence/absence of cultural materials and within site boundaries to mark the frequency distribution of artifacts of various sorts.

If they are used to identify and define artifact distributions, then excavated spoils may be screened through 3 mm or 6 mm ($^1/_8$-in or $^1/_4$-in) hardware cloth.

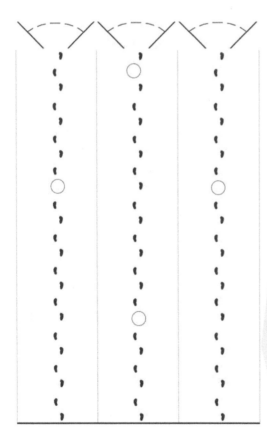

	Area Scanned
○	Shovel Test
'	Paces

Shovel Test Interval

STPs are dug at a fixed intervals, with the length of the interval depending on the intensity of coverage required to meet research needs or regional guidelines.

There are different ways to create an STP grid pattern. This diagram shows STPs excavated at a 15 pace interval (for this long-legged person, about 15 m/ 49.2 ft). Alternating the first STP at 15-7-15-7 paces creates a pattern that maximizes the systematic dispersal of test holes.

Surface Scrapes

Under some environmental conditions, surface scrapes are more useful than shovel test pits.

Surface Scrape Methods

Use a trowel to make the scrape, or carry a rake or long-handled hoe which can also serve as a walking stick.

Surface scrapes are appropriate in areas with poor to no surface visibility due to evergreen duff, leaf litter, or other loose material on the surface.

Scrape through the duff and well into the soil looking for cultural indicators like ceramics, glass shards, metal fragments, stone flakes, fire-cracked rocks, bones or mollusc shells, discolored soil—whatever indicates an archaeological site in the region.

On landforms where research indicates cultural materials are not likely to occur, scrape at broad intervals (20 to 50 m/65 to 165 ft).

On landforms where research indicates cultural materials are likely to occur, scrape at close intervals (5 to 10 m/16 to 33 ft).

Look for rodent disturbance, recent bulldozer or backhoe scars and spoils, or—best of all—deep stream cuts and banks in your transect line. Check these thoroughly for any sign of cultural material, buried cultural strata, or features like hearths or firepits, architectural remains, burials, or concentrations of rocks.

Adapting Methods to Purpose

Chapter 12

Eventually, any archaeological survey must get down on, and sometimes into, the ground to look for sites. The exact methods of search in any given case are dictated by the nature of the local environment and the intensity of survey required for the kind of planning being done. Five basic points should be kept in mind in planning field work:

- The field work should make maximum use of background information.

- The field team should include people trained to recognize all the types of archaeological phenomena that are likely to be there.

- It is often most effective to conduct the field work in several stages of increasing intensity.

- Field methods should be planned carefully to allow for environmental diversity.

- Within reason, all ground surfaces should be inspected and subsurface exploration should be done if the surface is obscured or if there's reason to think that buried sites may be present.

The Griffin Valley Example

To illustrate these points, let's return to Griffin Valley, and assume that we have done enough background research to know (a) the general outline of the area's history and (b) the results of the Beakey and Loumington surveys. For an area of this size and complexity, a multistage approach to field work is probably appropriate. The first stage involves consolidating and verifying the knowledge we've gained from background study while familiarizing ourselves with the character of the valley.

Knowing of Beakey's discoveries, we immediately set out to verify them and determine the present condition of the sites he recorded. Because our background research has indicated the general locations of the Indian massacre and the 1872 gold rush, we do a cursory inspection of these locations too. We're aware of Loumington's work, but having an idea of the constraints under which she worked we don't have much confidence in her results.

We begin the field work with four small-scale cursory inspections in different parts of the study area, which of course require us to drive or walk through the area in general, getting a feel for it. At the same time, we fly over the valley in a light plane to develop a further acquaintance with it and to observe directly the field scars first noted by Beakey—which we think we've been able to detect in aerial photographs we got from the Natural Resource Conservation Service.

We know from our background review of soil maps that there's evidence that a lake once existed in the valley, but no one has established the elevation of its shoreline—which, of course, is where we'd predict that old villages or campsites might be found. In consultation with a specialist in geomorphology (a geomorphologist or geoarchaeologist), we undertake as a second phase of survey: two walking transects across the valley, looking for the old shoreline while verifying the locations of plant communities we've identified from remote sensing data (Chapter 9). We make no pretense of seeking full ground coverage at this stage; we're developing a general understanding of the area's key archaeological resources.

Team members are deployed about 15 m (49.2 ft) apart. Thus deployed, the team, including the geomorphology specialist, makes two sweeps across the valley. In the process we discover the old village site at the closed spring but miss the mammoth kill because of the wide spacing of team members and the low visibility of the site (represented on the surface only by bone fragments in gopher runs). We fail to

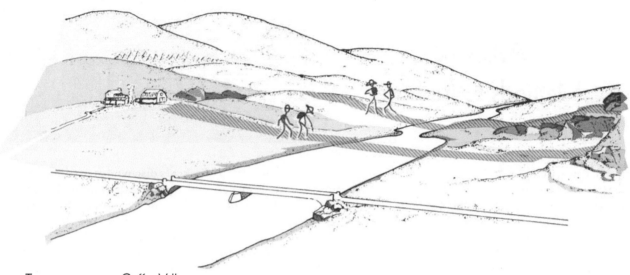

Transect survey in Griffin Valley.

identify the old shoreline but do get a good idea of the distribution of plant communities.

Armed now with a general sense of how the valley and its surroundings are organized and an enriched understanding of its archaeology, we plan for the next phase of work—an intensive inspection of the whole area.

Our team consists of eight people: an archaeologist specializing in prehistory, a colleague with expertise in historical archaeology, two experienced students, two members of the Ebirt Tribe, and two experienced members of the local avocational archaeological society. We've discussed the results of the background research and preliminary field work, so we have a pretty good idea of what kinds of prehistoric and historic sites are likely to occur and what they'll look like.

We form two teams of four people each and first attack the grasslands along the river. Because the ground surface is badly obscured by turf, subsurface exploration is necessary. We deploy team members in this case about 5 m (16.4 ft) apart. Each is armed with a small shovel. Each moves forward 20 paces, scanning the ground as he or she goes. Stopping, he or she digs a small hole, clearing the turf and penetrating perhaps 50 cm (1.6 ft) into the ground in search of flakes, pottery, artifacts, or other indicators of past human activity. When the shovel test is completed, the team member refills the hole and moves on another 20 paces.

Let us hasten to say that there's no magic in 5-m spacing, 20 paces, 50-cm shovel test pits, or the fact that our holes are more or less circular. Sometimes we space ourselves 3 m apart, 5 m, or 10 m. Some people prefer square holes. Sometimes it's necessary to dig them deeper than a half meter, or not necessary to dig so deep. Sometimes it's appropriate to put all the dirt through screens, other times not. It all depends on what we think it's likely we're going to find based on our research and experience in the area.

However it's done, shovel testing is a slow, expensive, frustrating, and often marginally effective way to locate archaeological sites. Small phenomena can still escape notice, falling in the cracks between team members. Individual team members will vary in their ability to see things or interpret them. Furthermore, shovel testing tends to discourage team members from inspecting their surroundings in general, forcing them instead to concentrate on pacing and digging. This can lead one to miss, say, the pictographs in the half-buried rock overhang, or the tree with a bow half-carved out of it. In addition, shovel testing probably creates a mental set that is less than effective as a stimulus to discovery; it can be really boring.

Many other methods of subsurface exploration have been and are being used by different archaeologists; they include the use of power and hand-driven posthole diggers, backhoes, tractor-drawn plows, hand-driven and powered cultivators, road graders and gradalls. Many of these techniques are obviously destructive, both of archaeological sites and of the natural environment, and they're also fairly expensive. Remote sensing, using technology like ground-penetrating radar, resistivity sensors, and magnetometers, isn't destructive but it may be even more costly, and skilled technicians may be in short supply in many areas. Consequently, shovel testing remains the most widely used technique for basic subsurface exploration.

Having surveyed the grasslands (if we're lucky, locating all the sites there) we can proceed to the chaparral zone on the south slope of the central hills. Here the ground is barren of grass and there's little soil development, so subsurface exploration isn't necessary. On the other hand, the survey is made more difficult by the chaparral itself which obscures the view all around and is almost impossible to get through. We have to deploy ourselves only 3 to 5 m (9.8 to 16.4 ft) apart and crawl through the dense brush on hands and knees.

Moving to the crest of the hills where there is little grass and no brush, we can take a break, drink some water, and then spread out at about 10-m (33-ft) intervals. We don't need to do shovel testing. We sweep rapidly east along the south side of the crest, record the palisaded village site, turn and sweep back to the west along the north side of the crest.

The oak and piñon forests offer special challenges; the ground is obscured by leaves in one case, needles in the other. Here team members deploy about 6 m (19.6 ft) apart and employ surface scrapes—a variant on shovel testing (Chapter 11). Armed with a rake, each team member walks 20 paces, clears the leaves or needles off an area about 3 m (9.8 ft) on a side, inspects it, then rakes the leaves or needles back into place.

Finally only the rocky slopes remain uninvestigated. In the rocks it's easy to lose track of direction and thus fail to look at some areas while looking repeatedly at others. To avoid this unhappy situation, we defined three survey transects and marked them with a series of control stakes with colored flags in the grasslands below the base of the slopes (illustration, this page). Deployed about 5 m (16.4 ft) apart, team members begin at the first stake and follow a compass bearing to the top of the rocky slope. Reaching the top, the team moves over until the reverse of their original bearing brings them down on the second stake, then they work their way down. Moving to the third stake, the sequence is repeated. The stakes are positioned about 15 m (50 ft) apart, so that with a four-person crew, there is a some overlap between transects. Shovel testing isn't necessary here, but the survey is very time consuming because of the

Survey of Griffin Valley rocky slope: areas scanned in Transects A, B, and C.

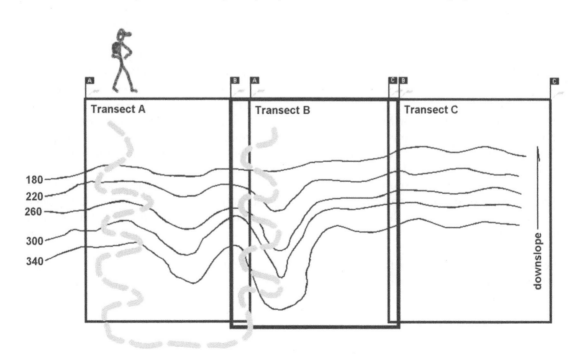

need to explore the complex rock outcrops for petroglyphs and pictographs, peering under overhangs and into cracks, staying alert for rattlesnakes.

While the basic surface survey is being completed, one small team undertakes a special study. During the cursory inspection at the beginning of the survey, we discovered that the vicinity of the Late Stoneland village site has changed considerably since Beakey's visit. In fact, because the area was converted in 1968 into a feed lot, we can't see any evidence of the village at all.

It appears that to create a stable, level surface for his feed lot, Mr. Ford graded down the low pass to the north and dumped fill along the river bank. Where Beakey reported a Late Stoneland village site we find, to our astonishment, a scatter of flakes, core tools, and fragmentary Clovis points. To reconstruct what happened, to determine the present condition of the Late Stoneland village site, and to delineate the boundaries of these complicated and much-obscured deposits, detailed subsurface testing is necessary. For this purpose, we use a backhoe to cut a series of carefully controlled trenches along the river bank. These trenches reveal a dense deposit of refuse containing Late Stoneland potsherds and projectile points at a depth of 2 m (6.5 ft). We then make minor cuts elsewhere in the feed lot (and promptly refill them) to define the north, west, and east boundaries of the site and to determine whether its burial in 1968 has badly disturbed it. At the same time, we carry out an intensive search of the feed lot to locate Clovis artifacts and waste material. Although obviously removed from their original context, the material at least documents the occupation of the area during Clovis times, and the types of artifacts and waste material present provide clues as to what the Clovis people were doing there. We interview Mr. Ford and his workmen to determine exactly where the material came from that was dumped on the river bank and to make certain that none of it was hauled in from distant sources. We then closely examine the source of the fill and make subsurface tests with augers and deep STPs to determine whether any remnant of the Clovis site remains in its original setting.

With the completion of the surface survey, subsurface assay, and detailed site recording, we are ready to prepare a definitive report on Griffin Valley's archaeological resources.

Control Your Methods, Don't Let Them Control You

No one should interpret our presentation of techniques applicable to Griffin Valley as a prescription for proper survey field work; it's merely an example. What is important is that in Griffin Valley (a) we have made maximum use of our background information; (b) the survey team drew upon specialists trained to recognize the particular phenomena likely to be present; (c) the field work was conducted in several stages of increasing intensity; (d) methods of inspection were carefully planned to allow for environmental diversity, and (e) *within reason*, all ground surfaces were inspected, with subsurface testing done where the surface was obscured and where buried sites were thought to be present. The exercise of professional judgment and experimentation with techniques are entirely appropriate in the development and modification of survey strategies.

The caveat about surface inspection "within reason" deserves some special attention. In some cases it is perfectly reasonable not to inspect the surface in detail. For example, slopes too steep for occupation, without rocks for rock art, and without caves or shelters or other attractants to or evidence of human activity are often and properly not subjected to detailed inspection. If a floodplain is covered with 20 feet of alluvium accumulated during the last 50 years, and a construction project being surveyed will disturb the ground only to a depth of five feet, it is clear that neither inspection of the surface nor subsurface exploration is necessary. If an area has been so badly disturbed that a site would not likely have survived, no inspection—or only cursory inspection to verify the disturbance—may be necessary. And if the Ebirt Tribe regards itself as descended from or responsible for the remains of the area's Clovis occupants, these remains may be important to them even if they have lost their archaeological research value. The determination that disturbance has been total, however, must be made by someone knowledgeable about local site types. In the case of the Griffin Valley Clovis site, even to know that a site once existed, and to have a collection of disaggregated material from it, may be vital to understanding the area's history.

Chapter 13

On their second transect through Griffin Valley our field team encountered remains of the second agricultural village (Chapter 9). The site was first recognized by the middle person on the transect, who—as she climbed the ridge and neared the crest—stepped into a ditch which she instantly recognized as a potential prehistoric feature. She called out the find to the other team members, who also crossed into and relayed news of the same ditch to the east and west, at approximately the same elevation interval. The team, alert to this feature and likelihood that it represented an archaeological site, then met and formed a plan.

It's time for team members to put all their skills to use and parts in motion. We've already decided that our survey would be approached most efficiently using three-member teams, and the real utility of this team structure is apparent now that we've found a site. As they meet and divy up responsibilities team members realize they have a lot to do. But, like any team faced with a big job, it's best to break the chaotic rush of responsibilities into small tasks. So, let's think about three goals, each with a set of tasks: assay, documentation, and quality control.

A Note on the Purpose of Site Assessment and Documentation

Knowing where sites are located and learning something about what they may contain is our obvious immediate goal. Why else would we conduct a survey? Well, finding sites and their contents is a reason to survey, not to document. Site documentation constitutes the categorical and systematic inventory of your professional observations at the time of the site visit. Thus, your assessment and documentation must serve two basic purposes: (1) to make a baseline record of the nature, extent, and variability of the resource, including its status and integrity and its archaeological content, structure, and variation; and (2) to produce

documentation that contains verifiable details to be used to assist periodic efforts to monitor site status and generate updates.

Most large projects involve at least some previously surveyed lands, so surveyors encounter a combination of new and previously recorded sites. In either case, documentation has the same purpose. If your team encounters a site that has never previously been visited or recorded by an archaeologist, then your record represents a new bench mark. If your team successfully reidentifies a previously recorded site, then your record should serve as an solid replacement, amendment, or supplement. Whether new or revisited, your site documentation should supply an assessment of new or ongoing impacts and an improved and updated account of verifiable details for future investigators who may or may not have ever visited the site before.

ASSAY

The Griffin Valley survey team's first goal is to assay the site, that is, define its nature, extent, and variability. The whole team will be engaged in activities related to this objective and will start by defining the content and horizontal boundaries of the site, establishing a mapping datum and permanent datum, scouring the ground surface to identify artifacts, and then taking steps to determine its subsurface extent.

Define Boundaries

Team members break up to scour different parts of the ridgetop attempting to define site boundaries. The first pass over the site results in the discovery of only six artifacts: a chipped slate hoe blade fragment, three coarse quartzite flakes, and two brownware pottery sherds. As we will soon determine, the site was only briefly occupied around A.D. 500–1000, so there are relatively few artifacts and these tend to occur clustered together in areas that mark former houses, earth ovens, and refuse heaps. Because our survey is taking place in the springtime and grass is high and thick on the ridgetop, the artifacts and features are very difficult to spot. Finding that surface inspection and even surface scrapes are inadequate to determine site content and boundaries, our team decides to set a grid of shovel test pits (STPs).

Having established a "datum" (a fixed position that serves as a reference point to identify, verify, plot, or coordinate) at the crest of the ridge, the STPs are done every seven paces (approximately 6 m/20 ft) in five alignments oriented north-south and spaced seven paces apart, for a total of 25 STPs in a systematic grid covering the ridgetop. All but two of these STPs are dug only to determine the presence or absence of cultural material, so each is dug in a single level to just 20 cm (7.9 in) deep, about half the length of the shovel blade. Two others are deep probes, described below. The STPs are numbered consecutively in an alpha-numeric grid, and their positions marked with pin flags in anticipation of plotting them on the site sketch map, also described below. The cultural materials observed in each STP are classified and quantified, and these details are recorded on a specially designed STP log form. Certain artifacts, including those that might tell us more about the age and nature of the occupation, are set aside for special treatment.

Map Datum and Permanent Datum

On an excavation the datum serves as the primary reference point of a sampling grid. On a survey, a datum often serves two purposes, and these purposes might well be served by two separate datums, a permanent datum and a mapping datum.

The permanent datum is a physical monument fixed on or adjacent to an archaeological site, and it is positioned and prepared expressly for the purpose of serving future reference needs. In most cases, the permanent datum serves as a verification point for periodic efforts to monitor site status. Because the permanent datum will be sought by investigators who may not have previously visited the site, it should be located near an obvious natural or cultural feature, like the base of a large, prominent boulder or the corner of a collapsed fireplace. If no such feature is apparent, then a location anywhere on or immediately adjacent to the

site will suffice, with the proviso that one should attempt to place the datum where it is least likely to be disturbed, at least in the short term. The permanent datum should be a fixed marker of some sort that contains a site verification code. Currently, it is common practice to sink a long stake of aluminum or other alloy, or fix a brass or aluminum cap to a boulder. A tap set is often used to punch in the site number (page 130). In the future, microchips, laser chips, bar codes, and miniature electronic signaling devices may be embedded in the datum to make it easier to find the datum and store essential site information.

In order to tie into Geographic Information System (GIS) computer mapping capabilities, you should make an effort to acquire reliable "x-y-z" (UTM/UPS northing, UTM/UPS easting, and elevation) coordinates on your datum. Acquiring this information, which is called "georeferencing the datum," is best accomplished using a high-quality receiver. However, we have acquired high-resolution positions by setting two or three 12-channel receivers side-by-side on the ground nudging the datum stake, waiting until all of the units acquire four or more strong satellite signals, then waiting until the difference between the positions registered is at its absolute minimum and averaging this difference or until all the units register the strongest possible signal strength, at which time no difference may be observed.

In contrast, the map datum is strictly temporary and is located strictly for immediate needs, serving as a reference point for mapping, plotting, photography, and other forms of site documentation. The position of the map datum is selected on the basis of its optimal lines-of-sight to all or most corners of the site and surroundings. Often, the map datum is not an object or natural feature, but simply a spot with clear views. Thus, don't feel compelled to find a useful object or feature (e.g., a stump). Simply drop your pack on a suitable spot and start drawing (page 134).

Some agencies, project directors, or crew chiefs will insist on using the permanent datum as the map datum. We've found this extremely inefficient in most cases because the permanent datum is by design set snugly against some obstruction (like a large boulder) that may block one's view to half the site or more, making it impossible to shoot bearings. We've also found this requirement silly because if you make the best possible map, all pertinent natural and cultural features—including the permanent datum—will be plotted accurately. Thus, it shouldn't matter where you establish your mapping datum, the real emphasis is on making a thorough and accurate map.

Flag, Tag, Bag, and Log Artifacts

Our Griffin Valley survey team has entered the field with a clear and concise collection strategy (Chapter 9). Now that we've found a site, one team member, the "artifact honcho" is assigned to take control of all artifact collection and documentation. The artifact honcho lists the artifacts on a specially designed artifact log form to be attached to the site record, adds the artifact number to the pin flag (or places a prenumbered flag), then pulls the artifact, fills out an artifact tag, and slips the tag into a collection bag with the artifact. By assigning one team member to the task of flagging, tagging, bagging, and tabulating artifacts we make it much easier to maintain a consistent and systematically applied numbering system and a thorough treatment of artifacts.

Remember, the same artifact number goes on the site sketch map, the artifact record attached to the site map, the pin flag, and the tag in the bag with the artifact.

The honcho must be very alert during this process—artifact numbering errors cannot be solved once you walk away from the site, and this kind of error instantly loses provenience, the fundamental starting point for archaeological chain-of-evidence. The authors have seen occasions when artifacts were left abandoned in the field, sitting on the ground, in bags with tags—a disheartening and avoidable mistake. Artifacts are very easy to misplace, and it is common to find an artifact and then—moments later—not be able to find the spot or artifact again. So, the best thing to do is to flag, bag, tag, and log artifacts immediately. This requires coordination with the full crew. Crew members engaged in site assay or other documentation tasks will continue to find artifacts as they work around the site, on the surface, in STPs, in surface scrapes, in creek banks, and so on. Every time someone finds an artifact they should call over the artifact honcho. All artifacts should be flagged immediately by the honcho, and artifacts to be collected should also be bagged, tagged, logged, and collected immediately.

Determine Subsurface Extent

For the purpose of site documentation, methods typically used to do subsurface probes—auger probes,

STPs, and creek bank inspection described in Chapter 11—are focused on the site and its immediate surroundings and used in this case to determine subsurface extent. Our Griffin Valley team has dug a few selected STPs to a maximum possible depth (usually an arm's or shovel handle's length) and has logged their stratigraphic observations, satisfying our immediate research design and field sampling objectives.

DOCUMENTATION

The assay is complete, and once again the team will gather together, this time to discuss what they observed. New tasks are also identified. Although the artifact honcho continues to flag, bag, tag, and log artifacts, the other two team members shift responsibilities, with one selecting a datum and beginning a sketch map and photo documentation and the other starting a site record. Observations made during the initial assay will now come into play for documentation—for example, the site map will depict site boundaries, artifact locations, and the STP grid, and the site record will describe the site location, list the artifacts, describe the soils and constituents, and provide an estimate of site depth and function.

Site Sketch Map

Site sketch maps and imagery have two purposes: (1) to depict the status of the site and the distribution of natural features, soil variability, constituents, artifacts, and features at the time of observation; and (2) to provide enough accurate landscape detail to enable subsequent visitors to relocate the site.

Selecting the Paper

Since our team has already selected a mapping datum we can now turn our attention to the site sketch map. The site sketch map begins and ends with the paper and grid.

Most government agencies issue standard site forms which often include a site sketch map form. You are really only required to use this form to submit your final, inked map. Generally, these forms are not useful in the field because they are often no more than a formatted frame lacking grid lines. So, it is best to use a sheet of graph paper or engineering paper to complete the map in the field. Back at the lab, the field sketch map can be scanned, inked, and stretched proportionally to match the frame and then printed on the form.

Except in rare circumstances, we recommend completing the site sketch map on an 8.5x11-inch sheet. This is because the final inked map will be nearly this size, and if you make your field sketch map on a larger format paper, say 11x14 inch, you will have a tendency to add too much detail at too fine a scale and thus produce a final map that does not reproduce well at the smaller size. Further, the smaller format is good because it will make you work a bit harder to select what is really important and place on the sketch map those natural and cultural features that convey the most about the site and the landform it occupies.

There are lots of kinds of graph paper, but we prefer a coarse rule on 40-lb bond with a metric grid of five squares to the centimeter, or an even more coarse English grid of five squares to the inch, because it is easy to convert distance measurements and because a fine-rule paper is too busy. Continuing on the "busy" theme, standard engineering graph paper is often inked on the back face and designed for a slight bleed so the grid lines appear faded on the front. Use the faded face to make the map. Otherwise, if you use the inked face the grid is so coarse and bold that the map can be hard to read.

Matching the Scale

Matching map scale to paper size is a crucial first step in making the sketch map. In essence you should, on the one hand, make optimum use of the paper by filling it with a reasonable amount of site detail and, on the other, include as much information about site surroundings as useful and feasible. These two functions emerge from the two purposes of the map, discussed above.

Our Griffin Valley team has taken steps to serve both these purposes by taking the time to decide the area to be included on the sketch map while examining the perimeter and establishing site boundaries. The mapper will make sure the map extends out far enough to show the general shape of the ridgetop on which the site resides.

In other situations, the mapper may decide on different solutions. We simply need to remember that the area selected for depiction should cover key natural features (e.g., lake shore or stream course) and significant constructed features (e.g., roads, streets, powerlines, structures) that convey site status and help others find the site.

Our Griffin Valley team has decided how much space they want to depict on the map. Now, they need to find the best place to position the datum on the page. To do this, the mapper first paces the maximum site length and width and next takes a quick bearing on these dimensions to determine the general layout of the site. Next, the mapper will fit the maximum dimension of the site to the maximum length of the paper. The Griffin Valley ridgetop site and perimeter landform is roughly 100 m (328 ft) north, 50 m (164 ft) south, 60 m east, and 40 m west of the mapping datum. So, the mapper has a relatively easy job, and makes the center line of the paper the north-south axis of the map and places the datum one-third from the bottom of the page.

Plotting the datum is rarely this easy. Generally, you will need to adjust left-right and top-bottom to best center the map on the page. Be aware that there is a common habit among fieldworkers of placing north at the top of the sketch map. In most cases, this habit is impractical and will only force the mapper to make a poorly proportioned map. Just to make it clear: the centerline of the page can be any bearing that makes best use of the page.

Back at the ridgetop site, our mapper has determined the best plot for the datum on her engineering-ruled, five-squares-to-the-inch graph paper. She now turns her attention to establishing the map scale. Since the map area should measure about 150 m (492 ft) north-south, and our mapper knows there are about 10 usable big squares on her sheet, she decides to make the map scale 1 in=20 m. Thus, every small square represents four meters. This gives her a usable map space of about 200 m north-south and 140 m east-west.

Because our mapper does not have a ruler in tenths English scale, she cuts a long strip off one sheet of the graph paper, and makes her own quick, custom ruler with marks at 10-m (½-in) intervals.

Map Key

Prior to field work, the Griffin Valley team settled on a standard key to be used to draw sketch maps for the duration of the project (page 131). The key is composed of standard symbols for natural or cultural features likely to be observed on many if not most sites identified in the project area. If a cultural or natural feature depicted on a site sketch map is represented on the standard key, then there is no reason to include the object on the individual sketch map key. Thus, you only need to use the individual sketch map to depict unique items.

Standard keys should be carefully designed to make each element relatively easy to draw and marked by a distinct appearance or aspect that readily evokes the feature depicted and enables the viewer to differentiate one thing from another on the sketch map. Render your own set or feel free to borrow the set we provide here.

Pacing and Plotting

Now it is a matter of shooting in bearings and pacing out notable natural and cultural feature. We like to shoot bearings about every 20° like the spokes on a wheel, eventually sweeping around the perimeter.

First, our mapper shoots a bearing on a cultural or natural feature to be plotted on the map, then paces the distance (page 106). Once the mapper reaches the feature, she consults the pace chart (pages 107–108) for the actual distance.

With bearing and distance in hand, she then places the see-through protractor on the graph paper, with the center aligned to the datum and N0°E at its proper place at true north (pages 134). She then finds the bearing on the protractor, puts a mark on the sheet, lines up the ruler on the datum and bearing mark, and measures out the actual distance to scale (page 135).

The most common mistake when plotting cultural and natural features is to inaccurately depict their scale in proportion to other features. The best way to solve this problem is to measure the area of the feature and depict it accurately. For example, if you are drawing a tree, shoot and pace to the trunk, but make a note of your pace interval when you pass under the tree umbrella. For a large tree, the trunk is generally at the center but the umbrella might cover an area of 15 m (49 ft) or more. Use your ruler to measure out the area contained under the umbrella and show this on the sketch map. Sketch that particular feature as you stand nearby and it will help you add verifiable detail.

There may be several cultural or natural features encountered along each bearing, and we like to keep a running tally on one edge of the sheet in light lead that can be easily erased. This tally might read:

    15 WO
    24 cutbank
    37 boulder
    54 WO
    72 site boundary

where the number is the pace interval followed by an abbreviation for a specific feature (e.g., WO=white oak). We like to take the time after pacing out each bearing to convert and plot each item then, as we go, we can begin to "connect the dots" on features intersected on several bearing (e.g., a cutbank).

The Griffin Valley mapper is filling in as she goes, and the sketch map is nearly done. In the final step, she checks various plots and proportions, adding detail on site status indicators or enhanced renderings that might assist with future site verification needs (page 135).

Sites that occupy a large area are often composed of several distinct loci and cultural features. Depending on your research design, purpose, or time and funds available, you might make detail maps of particularly important or interesting cultural features. Detail maps vary in scope and scale depending on the features encountered.

In Chapter 3, we recommend carrying two mechanical pencils filled with HB lead. Why two pencils? Using a pencil combo of say, 5 mm/9 mm, will allow you to draw more distinctive symbols and features on site sketch maps. We recommend using the large lead to make bold lines for natural features, and the smaller lead to make fine lines for topographic and cultural features.

Map Legend

Once the basic map is done, our Griffin Valley mapper adds a sketch map legend (page 134). The legend lists basic identification and documentation protocols, such as: (1) site name and number; (2) map scale (with measurement units); (3) a north arrow showing map orientation; (4) the date; (5) recording team member names; and, (6) the UTM position of the site mapping datum.

Site Imagery

Like other forms of site documentation, site photography and video capture has two purposes, verification for return visitors and documentation of site status. Imagery—when it is applied thoughtfully—can serve these purposes exceptionally well, and so it should be considered an essential component of site documentation.

The days of film-based imagery are over, and these days nearly all professional archaeologists go into the field equipped with digital cameras. Digital camcorders have also come into common use in recent years.

Imagery for the purpose of verification should include photographs or video of the site shot from likely nearby parking spots or lines of access, for example, taken from a trail or road leading to the site. Imagery for the purpose of verification should also capture matchable details like unusual trees, outcrops, or other distinctive features. Verification is also served by imagery shot from an overview vantage point or shot from the site datum toward some distinctive feature.

Imagery aimed at documenting site status might be very different, focused on individual artifacts and features, or aimed at ongoing problems in site integrity such as looting pits, creek bank erosion, or trail and road encroachment.

Our Griffin Valley site mapper uses a digital camcorder to take a full, 360° pan around site GVP-07-003 from the site permanent datum. She then switches to a digital camera and takes additional photos including site overviews, feature imagery, and context shots. She adds the photo stations to the site map using a small "C"-shaped symbol showing the station, photo number and bearing of the shot. The photos and video are logged using a photographic record form completed and filed with the full site record (see below).

Site Record

Selecting a Site Name or Other Identifier

Alpha-Numeric Codes. In the U.S., your site documentation will ultimately be filed with the state office of historic preservation, tribal historic preservation officer, and/or state archaeologist's office—each of which may review and comment on the quality and completeness of your documentation—and when accepted will be assigned a site trinomial, a numbering system developed by the Smithsonian Institution in the 1940s to support the river basin surveys (Chapter 2) and still in common use for standardized site numbers nationwide. In this system, a number designation is given to the state (01–50), a letter code to the county, and a number in sequence assigned to the site. There is a little bit of expectable variation in the way this system is applied nationwide. The original state number code was bumped when Alaska and Hawaii were added to the U.S. in the 1950s, but some states still stubbornly cling to their original number in abeyance of the change. Other states responded by replacing the number code with the more common state mailing acronym.

Nevertheless, the Smithsonian trinomial is ultimately assigned well after the conclusion of field work. At the same time, there is a need for careful inventory and tracking in the field, so most states' office of historic preservation and companion federal agencies distribute site record forms that allow for one or two additional names, usually a field-based code identifier and a formal name. Other countries use systems that—while often superficially quite different from the one used in the U.S., are conceptually quite similar. Sites are assigned sequential numbers within some sort of defined space like a province or administrative area. Sometimes the designation system includes codes to indicate the type of site represented or other pertinent data. Whatever the local system is, you need to learn and use it.

With respect to the field code identifier, if you are working for a particular state or federal agency, then you may be required to name the site using a specific management unit identifier. Alpha-numeric binomials, trinomials, or quadrinomials are common for this purpose. For example, the U.S. Forest Service uses a four-part numeral indicator identifying region, forest, district, and site.

However, absent this requirement, traditionally, on large projects site names are designed to fill project-specific inventory needs. For example, our Griffin Valley site recorder identified the hilltop settlement site as "Griffin Valley Project, 2007, Site 03" or GVP-07-03.

Appropriate Site Names:

- Based on a project name and number series (e.g., GVP-07-02)

- Based on a local landform (e.g., Griffin Creek Terrace Site)

- Based on an address or location (e.g., Indeterminapolis Market and 7th Site or Griffin Creek Campground Site)

- Based on a unique local ecological characteristic (e.g., Blue Falls Site, Buzzard Roost Rock Site, Sentinel Bluff Site).

- Based on a demonstrably associated historical period, person, place, or event.

Inappropriate Site Names:

- Based on crew member's or pet's name

- Based on personality traits.

- Based on project events.

- Based on personal beliefs or convictions.

- Based on romantic or maudlin sentiments.

- Based on hostile, disrespectful, or discriminatory phrases.

- Based on personal codes (e.g., hide a message in the acronym).

Some place names are inherently disrespectful and discriminatory, and you should avoid using these terms. You might also consider the interests of tribal groups or other stakeholders who may reject site names based on landowner family names historical families associated with a checkered, discriminatory, or violent past.

Formal Names. If you know that the site has a historical name (Sally Smith's Farm, Valhalla, etc.), this needs to be indicated on the record form, together with any uncertainties about it ("rumored to be Valhalla"). Otherwise, most regional archaeological traditions allow the recorder to call the site whatever he or she pleases. Some people use geographic references (the Red Lake Site) or take the opportunity to honor the site's owner (the Mohammad Khalid Site). Others get more creative (the Sunburn Site, the Fat Pig Site).

Particularly if you're tempted to become creative, please try to select a respectful name. Think about the future, when the site may be visited by the representative of a descendant community, members of a local historical society, or agency people who have the power to decide whether the site will be preserved, or an excavation funded. The name you select should leave an impression among all stakeholders that you have a deep and abiding respect for them, for the site, and for your own profession. The box on the left offers examples of appropriate and inappropriate site names.

Site Location

The site record should offer a variety of location information, referencing every reasonable and applicable legal, geographic, and coordinate system likely to be referenced by future visitors.

Map Location. The map location description should include information on the USGS quadrangle or equivalent—for example, the relevant Ordnance Survey (OS) map in the United Kingdom—covering the site, including the map name, map production series number, the date of production, date of photo revision (if applicable), map scale (in the U.S., 15' or 7.5'), and mapping datum of reference. Note that most site record forms also allow for a site location map (distinct from the sketch map) which consists of a precise plot of the site on a reproduction of a portion of the quad sheet or sheets covering the site and its surroundings. The location map should also provide a legend listing map information and a proportional scale. Consult Chapter 5 for a refresher on map name and identification information. Here is what our site recorder wrote down for the GVP-07-03 map location:

USGS 7.5' Griffin Peak, Indet. (1959, p.r. 1985), Map Series 3IV4/3VS, NAD83.

Coordinate Location. The site record should list site location in several coordinate systems including the state coordinate system (if applicable), and PLSS baseline and meridian, township and range, section, and ¼ of ¼ of ¼ section or other aliquot part. Latitude and longitude are the common reference system in some states and many countries outside the U.S. UTM zone and coordinates should be provided, at a minimum, for the site mapping and/or permanent datum, and the method of determination (map plot or GPS receiver) should be indicated. For very large sites UTM coordinates should be provided for points around the perimeter, and for linear features UTM coordinates should be listed for set intervals or otherwise plotted at significant changes in direction or elevation. Consult chapters 6 and 7 for a refresher on the PLSS, latitude and longitude, and the UTM system. Here is what our site recorder wrote down for the GVP-07-03 coordinate location:

> The site is located in the NW ¼ of the SW ¼ of the SW ¼ of Section 36, Township 4 North, Range 6 East, Mt. Halloween Baseline and Meridian. The permanent datum is situated at UTM coordinates Zone 5, 663543Easting, 4354678Northing, and at latitude 40°35'N, longitude 79°30'W.

Legal Location. At a minimum, the record should list state and county name; however, if the county assessor's parcel number is available or a specific municipal or rural address is available, this should also be listed.

Narrative Directions. Narrative directions provide road names and travel mileage to and from a specific headquarters (from an agency) or a major physical feature to a local embarkation point near the site. As for this embarkation point, mileage descriptions should deliver the reader to the closest possible appropriate pullout or parking spot, and at this point directions should switch to bearing and distance, with comments on passage around physical barriers or major waypoints along the path to the site. Here is what our site recorder wrote down for the GVP-07-03 narrative directions:

> From Indeterminate State Parks Division Headquarters in Indeterminapolis, travel west on Hwy 835 for a total of 13.75 miles (22.13 km) to a wide shoulder on the north side of the road. Park in the shoulder near the west bridge abutment near milepost 84.5. On a bearing of N10°E, proceed 250 m on the west bank and parallel to Phillips River, across two small ravines and then to the top

of a low, rounded, wooded hill. The site datum is located on the summit of the hill at the base of a large (45"BHD) red oak.

Narrative Site Location. The narrative description of site location is qualitatively different from other kinds of description, conveying instead information about the natural context of the site and a description of its landform and general geomorphic setting. The narrative description should be graduated, or nested, with the narrative carrying from the spot, to the locale, and on to the district, region, and area. The first descriptive phrase should describe the landform containing the site, and the last, the quadrant of the county. Here is what our site recorder wrote down for the GVP-07-03 narrative description:

> The site resides on the summit of a low, rounded hilltop overlooking the west bank of Phillips River, 300 m north of its confluence with Dry Creek, on the west side of Griffin Valley, 3.5 miles south of the town of Jacksalmon, in northwestern Indeterminate County.

Site Description

The site record should contain a thorough description of the nature, extent, and variability of the archaeological deposit. The description should contain a summary statement of the likely site type, followed by an account of the evidence including: site area, likely depth, soils and constituents, and artifacts and features observed. Here is what our site recorder wrote down for the GVP-07-03 site description:

> The site is a large hilltop settlement dating to the Late Archaic Period. The site is roughly elliptical in plan view and occupies an area of 150 m north-south by 85 m east-west. Shovel test pits indicate site depth exceeds 75 cm, but maximum depth is unknown. A total of seven features were observed, including: (1) a perimeter trench probably representing the remnant of a defensive earth work and palisade encircling the entire site. The perimeter trench exhibits two breaks representing probable gates or passages, one measuring 4 m wide on the west margin and one measuring 3.7 m wide on the north margin; (2) a small mound measuring 13x8 m on the east margin of the site at the break of slope and partly lagging into the perimeter trench. A shovel test pit in this locality indicated that the heap represents a trash midden

containing approximately 343 pottery sherds per m³, along with artiodactyl bone, charred maygrass seeds, charred acorns, charred squash seeds, fire-affected rock, and broken groundstone; (3) three clusters of fire-affected rock probably representing former exterior hearths or earth ovens. The rocks are composed of fractured and broken local field stone of metasedimentary material and a sample of ten from one feature measured on average 15x12x10 cm; (4) two depressions representing probable pithouse remnants. The depressions are roughly ovoid in plan view, shallow conical in cross-section, and both are marked by low perimeter berms. The depressions are similar in size and shape and (including berms) measure 8.5 m north-south by 6.5 m east-west and are a maximum of 35 cm below grade. The site is marked by a moderately dense scatter of pot sherds including cord and fabric-impressed and incised and punctate-decorated wall fragments. Artifacts observed included one small fragment of carved mica, one small fragment of carved pipestone, two grooved axes, eight nutting stones, four quartzite cores and numerous quartzite flakes, and 13 metasedimentary mano and metate fragments. Six small corner-notched and barbed arrow point fragments of varicolored flint were also observed.

Other Records and Observations

Site records should also provide a careful and complete enumeration of all observed features and artifacts, a tabulation of STP recovery listed by pit and depth, and an inventory of imagery taken during your recording visit. Note also that the feature tabulation (alpha) and artifact tabulation (numeric) should be consistent throughout the documentation. Artifacts, features, and photo stations should be plotted on the site sketch map. On large, complex sites, detail maps showing feature layout and artifact locations may be necessary. Make sure the insets are properly labeled on the master sketch map and the detail maps included in the final site record. An example site record form appears in Appendix B.

QUALITY CONTROL

Each team member will be assigned a series of tasks, and in order to bring some sort of reasonable flow to the effort, each team member should plan to work efficiently and cooperatively, the latter because each

team member depends on all others to accomplish several complicated tasks. It's important for each crew member to do thorough work the first time through, because a solid effort now will alleviate the need to return later to nudge documentation or do things forgotten or not done well the first time.

Because of this need for efficiency there is a tendency among crew chiefs to allot documentation tasks among team members according to natural and learned talents, and indeed there are field professionals out there—known for their amazing and evocative site sketch maps, for example—who should be tapped for their special talents at every opportunity. For example, when it's time to document a site one of the jobs is a natural for a good illustrator, the other for a descriptive writer, and so on. However, we would urge students not to succumb to specialization and—especially when you are learning—to get practical experience in all documentation tasks by rotating roles every time you find a site. You should seek to become proficient in all tasks.

For example, one of the most common errors found on site records is a conflation of site location and description, that is, a mixing of these two different kinds of information such that the description includes a location narrative, or the location narrative contains site description information. Going back over the documentation to get this straight and providing feedback to the team will help improve the quality of documentation with each site visited and help further define and differentiate the roles and responsibilities held by team members.

If there is some sort of problem and we need to correct an error, then we definitely will do so and be honest and open about it to all parties. However, we need to keep in mind that returning to the project area and site will probably strain the budget, possibly impinge on our ability to meet permit specifications, and stretch our polite adherence to landowner requests. So, we need to take the time in the field, before we walk away, to make sure the work is solid and that none of the essential documentation has been forgotten and nothing has been misplaced or done poorly. All this requires some quality control before the team leaves the site, and this is best accomplished by a crew chief who has been trained to spot and correct shortcomings should any occur. Crew chief tracking forms useful for quality control appear in Appendix C.

Steps to Take Once You Find a Site

Field crews vary widely in size, but a full crew is usually broken into individual teams assigned specific tasks. A three-person team is ideal for survey and its efficiency is especially evident when it's time to record a site.

Identify
When a crew person first identifies potential cultural features or material, other team members should be called over.

Define
The first task—accomplished by full team fanning out from the point of discovery—is to determine the nature and extent of the site. Team members engaged in this task will try to identify additional cultural material and define site boundaries.

Assay
Based on the density and distribution of surface finds and on the assessment of landform and soil type, subsurface probes may be necessary to determine the nature and extent of the site.

Document
Once the nature and surface extent of the deposit has been determined and subsurface probes completed as necessary, then tasks related to documenting the site can begin.

A "datum" is a fixed position used as a reference point to identify, verify, plot, or coordinate. Datums have two functions, often served by setting-up two separate datums—a permanent datum and a map datum

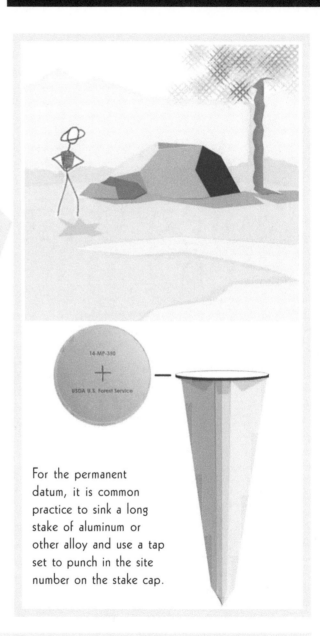

Select a Datum

Permanent Datum: Long-Term Verification

The permanent datum is a physical monument installed by an archaeologist. It serves as a verification point for later site visits or reference point for periodic efforts to monitor site status.

Because the permanent datum will be sought by future investigators who may or may not have previously visited the site, it should be located at or near an obvious natural or cultural feature like a prominent boulder, the corner of a collapsed fireplace, etc. If no such feature is apparent, then any location on or immediately adjacent to the site will do, but try to pick a place where it is least likely to be disturbed.

Map Datum: Immediate Lines-of-Sight

The map datum is a temporary reference point for immediate mapping, plotting, photography, and other documentation needs. The map datum should be on the site, at a location selected based on its optimal lines-of-sight to all or most corners of the site and its surroundings.

For the permanent datum, it is common practice to sink a long stake of aluminum or other alloy and use a tap set to punch in the site number on the stake cap.

A natural or cultural feature will often make a poor map datum location because the feature itself can be an obstacle to pacing or may block your view of one-half the site or more. If the permanent datum is less than optimal, or if no other appropriate physical feature is available, simply drop your pack or vest or stick in a bright pin flag at the best spot and start mapping.

Artifacts

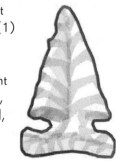

It will help to enter the field with a clear and precise artifact collection policy that is (1) linked to your research design; (2) generated in consultation with pertinent agency, landowner, tribal, or historical interests; and, (3) grounded in a curation plan.

On discovery, the crew chief should assign one person the task of identifying, flagging, tagging, logging all artifacts and features observed, and bagging all artifacts designated for collection.

Use pre-numbered pinflags to mark artifact locations. If possible, place the pin flag so it is visible from the site mapping datum.

Pin flags come in many colors, and you can use different colors for different artifact classes, or artifacts versus features.

When you are done, pull the flags and use them again at the next site.

Make sure the flag, tag, bag, log, and sketch map all list consistent site number, artifact number, and descriptive information.

The map key identifies the symbols used to make a map. Adopt a standard sketch map key, use and amend it according to project needs.

Standard Map Key

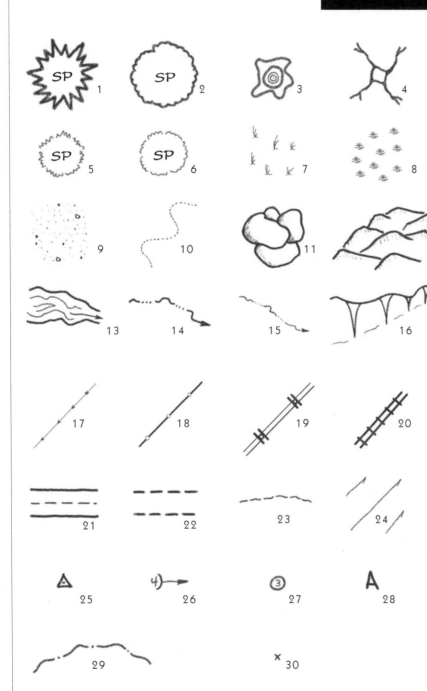

#	Description
1	Evergreen (species)
2	Deciduous (species)
3	Stump
4	Snag
5	Needleaf shrub (sp.)
6	Broadleaf shrub (sp.)
7	Grassland
8	Marsh
9	Barren
10	Soil or habitat boundary
11	Boulder
12	Outcrop
13	Perennial stream
14	Intermittent stream
15	Dry wash
16	Cutbank
17	Wire fence
18	Rail fence
19	Power/Telephone lines
20	Rail lines
21	Paved road
22	Dirt road
23	Trail or track
24	Slope indicators
25	Datum
26	Photo station
27	Artifact (recorded)
28	Feature
29	Site boundary
30	Artifact (observed)

Reproduce your standard key in an appendix or frontispiece to the site record volume of your final survey report. That way, there will be no need to repeat the standards in the legend for each sketch map. Reserve individual legends for unusual and unique features.

Standard key items should not anticipate everything, simply the most common things—those natural and cultural features, observations, and recorded elements likely to occur on or around most sites.

Shooting, Pacing, and Plotting

Once you've matched map scale to paper size and plotted your datum carefully to make maximum use of your graph paper, it's time to make the map.

First: Establish a datum point and select and shoot a bearing to a specific feature. This boulder is at N173°E.

Second: Pace to the feature, and convert paces to meters using the Pace Chart (Chapter 11). The boulder is 125 paces, or 121 meters from datum (at a pace rate of 0.85).

Third: Plot the bearing by centering the protractor on the datum and aligning it correctly to map zero and making a light pencil mark on the feature bearing.

Map scales vary, and you'll need different sized scales for different scale maps. Keep several rulers in your vest or purchase a multiscale ruler.

Shooting, Pacing, and Plotting (continued)

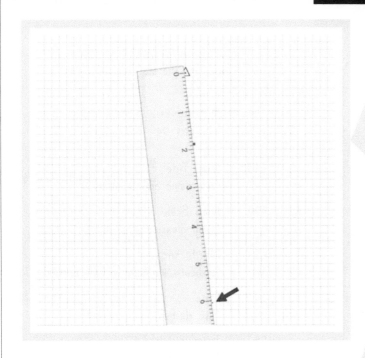

Fourth: Align a ruler from the datum to the mark, with zero on the datum. Measure out to scale and plot the feature at the correct distance. In this example, the ruler is marked in tenths of an inch, 1 inch = 20 m.

Fifth: Now plot the feature using an element from your standard map key. Remember to correctly proportion the feature by measuring its size or circumference and mapping it to scale.

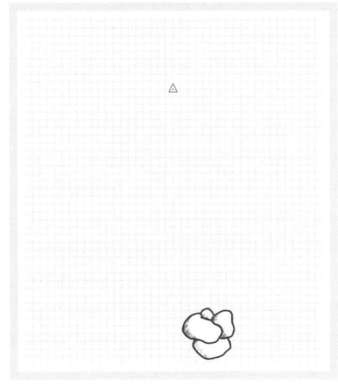

Fill in as you go, sweeping around the site in 45° arcs until the map depicts the complete site area and as much of its immediate surroundings as useful and feasible.

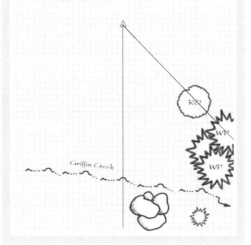

The map legend identifies the location depicted, scale and orientation, and important documentation background.

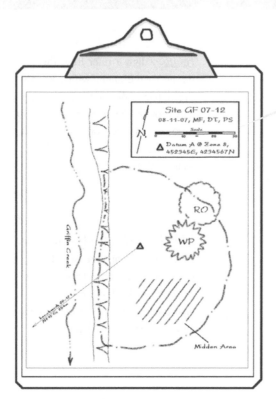

Leave one corner of your map open for the map legend.

The sketch map legend should always contain these five basic elements:

(1)   Map Title
      (site name and number)

(2)   Map Scale
      (with measurement units)

(3)   North Arrow
      (indicate magnetic or true north)

(4)   Date
      (use dashes)

(5)   Team Initials
      (mapper first)

(6)   UTM Position of Datum

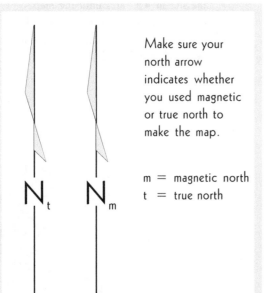

Make sure your north arrow indicates whether you used magnetic or true north to make the map.

m = magnetic north
t = true north

N_t N_m

Handwritten slash-separated dates are often hard to read because the slash can look like a number:

08/04/07

Dash-separated or full text dates are easiest to read:

08-04-07
August 4, 2007

Professionalism

Chapter

14: Being Responsible

Being Responsible

Chapter 14

As we finish our survey—and this book, let's consider a couple of issues that have to do with responsibility: responsibility on the one hand to people who are NOT archaeologists, and responsibility on the other hand to those who are, and may refer to our work in the future—as well as to archaeology and archaeological sites themselves.

REPORTING SURVEY RESULTS

Having been honest with ourselves about what kind of survey we've actually done and how we've done it, we need to make sure that what we've done, and our results, are properly recorded for future reference and use. As discussed earlier, one of the significant deficiencies of the data on file about Griffin Valley prior to our survey was that there was no way to determine how they were gathered. We knew from Beakey's survey that there were at least two sites in the valley, but we had no way of knowing whether his survey had been so detailed as to eliminate the likelihood that there were additional sites or whether, as in fact was the case, he had merely informally visited two sites and recorded them. In recording and reporting an archaeological survey it is vital to avoid this deficiency. This requires following a simple rule: Report exactly what was done and why and identify any uncertainties.

Reporting the Research Design and Plan

An archaeological survey report should describe the research design that guided the work, including operational definitions as to what was worth recording. Reasons for selecting the design should be discussed. For small projects, reference to a readily available regional or statewide design may be sufficient. The report should also discuss how the research design was translated into an actual survey plan—i.e., what the design meant to the archaeologist in the field.

Reporting Survey Methods

Early in any survey report, the methods employed in both background research and field work should be discussed. In many cases these may be separated into "background research" and "field work" chapters of the report. However it is done, it is important to report:

- What kinds of background data were thought to be needed, and what methods were used to find and consult them.

- What sources of background data were actually consulted.

- What difficulties, if any, were encountered in background research. What changes did these occasion in the research plan?

- What new or unexpected sources were discovered. What changes did they require in the research plan?

- What methods were employed in the field to search for sites. These should be described in sufficient detail to permit the reader to understand them fully and to appreciate the reasons for employing them.

- What variations among methods occurred at different phases of the survey or in different parts of the study area.

- When archaeological sites were discovered, what methods were used to define their boundaries and internal composition, to determine what categories of data they were likely to contain, and to define their significance.

- What areas were examined with negative results.

The reasons for choosing one method over another should be clearly explained. Portions of the study area where different methods were used should be indicated on maps, as shown in Chapter 10. In most cases, it is not necessary to report exactly where each team member walked or dug a shovel test, but it is possible to set modern GPS units to log individual transect waypoints and for crew members to take the time to record UTM coordinates for each STP dug. At a minimum, it should be possible for the reader of the report to reconstruct which methods were used in any given portion of the study area, to understand what these methods meant in terms of such factors as team deployment and subsurface exploration, and to understand the reasoning that went into selecting the methods employed.

Reporting Survey Results

Reporting Sites

Most states and many federal agencies, universities, museums and archaeological organizations use standard forms for recording sites. Generally speaking these should be used—with a couple of caveats. First, there are occasions when a standard form may not call for recording enough data or the right kinds of data to satisfy a particular need. For CRM purposes in the United States, it's usually important to record enough information to allow readers to judge the eligibility of recorded sites for the National Register of Historic Places. If the standard forms don't provide for recording this kind of information, or if the research design applicable to the survey requires additional data for some other purpose, the standard forms should be supplemented.

On the other hand, there are times when the standard form may demand too much data or the wrong kinds of data. For instance, an Indian tribe or other community associated with a site may not want a particular body of information recorded—perhaps because it's feared that others will misuse the data, perhaps because the data themselves are viewed as having some sort of spiritual power, or perhaps for some other reason. In such a case it may be important NOT to use, or at least not to complete, a standard form—instead presenting such data as can be presented in some agreed-upon form, and discussing why other information can't be presented. It is generally helpful to the reader of the report to summarize the form-recorded data in the text of the report. If there is the probability that describing the sites in detail in the report or providing their exact locations might lead to their destruction or damage by vandals or treasure seekers, the report itself may present only summary data with detailed information provided separately to those who need it for purposes of eligibility determination and planning.

The results of test or other excavations and of any special analyses conducted, should be reported. If collections of archaeological material were obtained, their depository should be identified, as should the depository of original field notes and associated data.

Reporting Other Discoveries

Discoveries that are pertinent to archaeology and historic preservation, but are not archaeological sites per se, should also be reported. Examples of such discoveries include but are not limited to: properties of possible architectural, cultural, or historical importance that apparently do not contain archaeological data; geological and geomorphological features that may bear on local paleoenvironmental studies; relict plant communities, pack-rat middens, and other biological features that may be of assistance in paleo-environmental studies; and, very recent cultural properties that may in the future be recognized as historically or culturally significant. No one should be discouraged from reporting field information that might lead to the discovery of a previously unknown historic property. Data that are of a proprietary nature and do not directly describe archaeological sites or other historic properties (e.g., proprietary information on geology received in confidence from a mineral exploration company) should not be reported without permission of the owner.

Reporting Areas of Uncertainty

If there are portions of the study area that appear likely to contain archaeological sites which could not be identified (e.g., places where deep alluvium, very thick brush, or modern construction made it impossible to inspect a location where background research suggests the likelihood that archaeological sites are present), these locations should be identified in the report. The reasons for uncertainty about their archaeological potential should also be noted and if there are reasonable means of resolving this uncertainty through further work, they should be presented.

Reporting Conclusions

Whatever conclusions we've reached about the archaeological resources of our study area need to be presented clearly, together with the thinking that underlies our conclusions. For example: "We found nothing over 1,000 years old, but observed deep alluvial deposits at the locations shown on Map 17B, which could easily have obscured older sites. Accordingly, we cannot say that the valley was not occupied prior to 1,000 A.C.E." We need to make it clear to readers of our reports not only what we concluded, but why we concluded what we concluded.

Where our survey was done for purposes of land management, planning, or compliance with historic preservation or environmental laws, we need to cast our conclusions in language that's relevant to the management or legal requirements involved. For instance, where our survey was conducted to gather data for a federal agency's compliance with Section 106 of the U.S. National Historic Preservation Act, we usually need to relate our conclusions somehow to the eligibility of places for inclusion in the National Register of Historic Places. This typically (but not always) means offering and justifying opinions about whether a given site, group of sites, landscape containing sites, or other piece of the environment is eligible for the register with reference to the criteria, standards, and guidelines set forth in pertinent federal regulations. The pertinent regulations and guidelines can be accessed through the websites of the National Park Service or the Advisory Council on Historic Preservation. URLs for these websites are listed below.

Our survey should have generated some information—besides simply the locations of sites and what they look like—that's of value in understanding local history or prehistory. Conclusions concerning local or general research problems in anthropology, history, or other sciences and humanities should be presented. Any local or other public interests that have been identified in the historic properties of the area should be discussed.

Keeping Track of Field Operations

High-quality reporting demands a high level of control over the nature of field operations. This means fairly detailed record keeping. Several examples of forms used to keep track of field survey data are given in Appendix C. Developing systems for keeping track of survey data is an important part of pre-survey planning. The exact type of system employed will vary with the nature of the project and the survey area. The forms in Appendix C may provide useful ideas and examples, but be prepared to adapt to your particular project circumstances.

BEING CLEAR ABOUT WHAT WE'VE DONE: SURVEY CLASSIFICATIONS AND THEIR DANGERS

Oddly, archaeologists—who almost by reflex divide pots, spear points, and cultures into named categories—have never established a standard system for describing surveys. Or to be more accurate, archaeologists have devised a plethora of systems. Over the last few decades in many parts of the country and, we suppose, the world, people—most of them archaeologists—have come up with survey classification systems. In one place we have "Phase I, Phase II, and Phase III" surveys; in another "Class 1, Class 2, Class 3 and Class 4." Often these classes and phases are broken down into subcategories—Phase IA, Class 2b. People use these terms, even in contract documents, as though everyone knew what they meant, usually without giving much thought to the implications of their use. But everyone does *not* know what they mean and—more importantly—different people think they mean different things, which can be pretty confusing.

In some places, for instance, a "Phase II survey" is defined as something like "a survey designed to gather the information necessary to evaluate the significance of all cultural resources located in the project area." Several assumptions are embedded in this terminology, including:

- some kind of definition of "cultural resources"—that is, a shared (though perhaps not widely shared) understanding of what the surveyors are supposed to be looking for;

- the assumption that it is possible to find all such things, whatever they are;

- a shared understanding of how "significance" is to be measured;

- that "the project area" is the appropriate focus of investigation; and

- that the methods needed to identify "all cultural resources" have been agreed upon.

We doubt if these assumptions are ever entirely justified.

If one defines "cultural resource" to mean only "archaeological site," it might be possible to be sure one has identified them all if one is surveying, say, an area made up entirely of bedrock where everything is visible on the surface. If one is surveying any place where there is soil deposition, obscuring vegetation, or pavement, however, one really can't be sure. And if one defines "cultural resource" more broadly (as most of us at least say we do, even if in practice we don't)— to include old buildings, structures, neighborhoods, villages and landscapes, places associated with a community's traditions and cultural practices, places where historical events took place, historical documents, and religious or cultural beliefs—one has even less chance of identifying everything. And most survey standards that we've seen guarantee that the surveyor will *not* identify everything, except the archaeological sites that their methods are designed to reveal.

The automatic assumption that "the project area"—usually taken to mean the location to be physically disturbed by a construction project—is the appropriate geographic area to be surveyed is occasionally justified but very often is not. It's a rare project that does not have effects of some kind beyond where the bulldozers roam—particularly if effects are understood to go beyond the physical to include things like visual and auditory impacts, and the "cultural resources" subject to effect are understood to be more than archaeological sites.

Finally, in part because of such ambiguities, there is seldom widespread agreement about just what methods are necessary to achieve complete identification of "cultural resources," or even archaeological sites. Must we dig holes every five meters, or only every ten, or not at all? Should we interview people who have lived on the land? Should we bring in experts in architecture, ethnography, geomorphology, botany? What training do our surveyors need to have? We can, of course, produce guidelines for training and practice, but the more specific these guidelines, the less likely they are to be widely applicable, and the more likely they are to mislead, or to lead to overkill.

When an archaeologist and whoever is contracting with him or her assumes that by doing a "Phase II" (or "Class I," or "Phase 3b") survey to identify "cultural resources" in a "project area," they have certainly found everything of historical and cultural importance, everything that's entitled to consideration under the law or as a matter of responsible practice, they are almost certainly mistaken. The results of this mistake can be both costly and destructive. Important archaeological sites are missed and destroyed, other types of cultural resources aren't considered in planning at all. People concerned with such resources aren't given the respect they deserve. Money is spent unnecessarily doing more archaeological survey than a given project merits, and important projects are delayed by last-minute discoveries.

We particularly hope that professional practitioners will pay attention to our attempt in Chapter 10 to classify archaeological surveys into types based on methods employed, things sought, and intensity of inspection. We hope this for a more serious reason than the mere desire to impose "our" system on others. We ought to be more careful about our terminology. We ought to be sure we know, and share our understanding of, what it is we're looking for and how we're looking for it. And we ought to understand, and share our understanding of, the real-world limitations on the accuracy and completeness of the data our surveys produce. The core purpose of this book, as it pertains to the work and thinking of professional archaeologists, is to encourage this kind of honesty and rigor.

AVOIDING MISREPRESENTATION

It's not uncommon for archaeologists working in cultural resource management (CRM) to refer to their surveys as "cultural resource" surveys. This is really a misrepresentation of what they've done—no doubt an innocent misrepresentation, but a misrepresentation nonetheless. It can cause "cultural resources" that are NOT archaeological sites—such as historic buildings, cultural landscapes, places important to people's religious practices, culturally important plants and animals, a whole host of things—to be ignored in planning construction and land-use projects. This undermines the purposes of the cultural resource laws and injures the interests of people to whom those other kinds of resources are important.

This book is about doing archaeological survey, not "cultural resource" survey. Identifying other kinds of cultural resources requires other kinds of methods, other types of expertise, and their comprehensive treatment would require a longer, rather different kind of book. In fact, for the purposes of CRM we shouldn't be conducting or requiring the conduct of archaeological surveys per se. We ought to be looking for all kinds of cultural resources, or at least all kinds of historic properties, using the interdisciplinary expertise and methods appropriate to so doing. We think that the continued division of the CRM world into disciplinary hege-monies, each with its specialized methods, is irresponsible, counter-productive, and silly. We hope one day to develop or contribute to a manual on the methods and uses of comprehensive cultural resource identification, but this manual is not it.

> ## Archaeological Ethics Websites
>
> **American Cultural Resource Association (ACRA)**
> *Code of Ethics*
> www.acra-crm.org%2FEthics.html
>
> **Society for American Archaeology (SAA)**
> *Principles of Archaeological Ethics*
> www.saa.org/ABOUTSAA/COMMITTEES/ethics/principles.html
>
> **European Association of Archaeologists (EAA)**
> *Codes of Practise and Conduct*
> www.e-a-a.org/eaacodes.htm
>
> **Register of Professional Archaeologists (RPA)**
> *Codes of Conduct*
> www.rpanet.org/conduct.htm
>
> **World Archaeological Congress (WAC)**
> *Code of Ethics*
> www.wac.uct.ac.za/archive/content/ethics.html
>
> **Canadian Archaeological Association (CAA)**
> *Principles of Ethical Conduct*
> www.canadianarchaeology.com/home.lasso
>
> **Archaeological Institute of America (AIA)**
> *Codes of Ethics and Professional Standards*
> www.archaeological.org/pdfs/AIA_Code_of_EthicsA5S.pdf
> www.archaeological.org/pdfs/AIA_Code_of_Professional_StandardsA5S.pdf

STAYING CURRENT

Archaeology is changing. New technologies are constantly being developed and introduced, especially terrestrial, aerial, and satellite-based remote sensing equipment and hand-held field devices used to record, datalog, and provide digital interfaces with lab and office-based equipment. These innovations will continue to alter how we conduct and track field survey, and the kind of records we produce.

In order to keep up you'll need to read the journals, newsletters, and websites of international, national, and regional archaeological organizations, attend meetings, and participate in forums on-line or in person. Often, practice is advanced well beyond the moribund standards that a state or federal agency may attach to your work. Always be prepared to advocate experimentation with new methodologies, if only to provide independent field tests in real world situations.

That said, it's necessary to pay close attention to whatever standards and guidelines have been developed in the state or region where we're conducting our survey. Often, specific practices pertinent to your project will also be spelled out as a condition of your permits, licensing, contracts, or report filing protocols. Even if we don't like them or think that they're inconsistent with law, regulation, or basic good practice as outlined in this book or elsewhere, we have to know enough about them at least to effectively adjust them.

Most state historic preservation offices (SHPOs) have issued survey standards and guidelines—sometimes quite rigid and detailed. Some SHPOs and agencies insist that these standards be followed to the letter, while others are more relaxed about them. It's important to remember that while there's no federal law or regulation requiring that a SHPO's standards be followed, compliance with federal regulations requires a "reasonable and good faith effort" to identify all kinds of historic properties, including but not limited to archaeological sites. If following a SHPO's standards will help meet this requirement, they should be followed; if not, then we need to adjust the standards to meet the actual legal requirement.

Federal agencies such as the Bureau of Land Management, Department of Defense, National Park Service, and U.S. Forest Service all recommend or insist on distinct methods and applications. Reflecting the longevity and robust nature of the U.S. highway archaeological salvage program, in many states the most elaborate and heavily reviewed and enforced archaeological standards and guidelines are found in states' department of transportation. Each state also issues guidelines, most of which have changed dramatically in recent years and will change again in years to come. The Council of Texas Archaeologists has posted an analysis of archaeological survey standards in 17 U.S. States, available on their website at http://www.counciloftexasarcheologists.org.

A web search should enable you to home in on the current standards applicable to your project. The table on the facing page lists websites with navigation tools, links or clickable databases for broadly applicable agency standards. Appendix A provides links to the archaeological survey standards, site identification conventions, recording forms, and archaeological survey report guidelines for all 50 states, two territories, and the District of Columbia.

Talking, and Working, with People

From the beginnings of CRM, archaeology has had a dominant role in cultural resource investigations, but this is also changing as a result new law, regulation, policy, and practice which gives a bigger role in survey and planning to other cultural resource concerns. This is especially true of the increasing role of descendant communities in all phases of archaeological survey. This will change how we—as a team—do prefield research and therefore how we—as a team—design our survey strategies. It will affect the kinds of resources we look for, the kinds of cultural traces we identify, and how we determine their nature and extent. It will change the kinds of information we gather and use to

record sites. It will also change the way we interpret and manage the archaeological record. In other words, it will fundamentally change archaeology and archaeologists.

One way to make our survey a little broader, and improve our rate of archaeological site discovery as well, is simply to talk with people. We've mentioned the need to do this at various points in the preceding chapters, but it's something that many survey projects don't do very well, and that's a problem. The need to understand people's perceptions of their cultural environment is especially evident when one gets beyond archaeology to deal with traditional cultural properties (King 2003; Parker and King 1990) and other places whose value lies more or less exclusively in people's heads. For example, in Western societies, a cenotaph represents a remembrance for the dead whose remains lie far away or may have never been found, and thus, people may visit, honor, and consecrate a place that lacks a direct connection or physical evidence. For some people, especially for descendant communities, a landform like a peak or lake, the waters of an entire watershed, or a particular plant species might have spiritual importance. These places may not have physical remains that can be detected using conventional archaeological methods. However, most kinds of cultural resource inventory must account for these places.

Cultural values exist in strictly archaeological contexts, too, and on a cultural resource survey the cultural values associated with an archaeological site must also be understood. The descendants of the people who created the archaeological sites—and those who, while not demonstrable descendants, nevertheless feel responsible for the places—need to be consulted so their concerns can be respected. The artifact collectors who have seen and often picked or dug up artifacts from a site are likely to have a far more intimate knowledge of it than any archaeologist can get in a quick inspection. Farmers, ranchers, loggers, forest managers—all may have relevant data about sites and their environments and important ideas about how both should be managed. And they may care about places in the landscape that—while not archaeological sites at all—are nevertheless important cultural resources that ought to be attended to in planning. We can't emphasize it enough: archaeological surveys should include finding and talking with real live people who know or care about the area being surveyed.

▷ **Secretary of the Interior's Standards and Guidelines for Archeology and Historic Preservation**
http://www.cr.nps.gov/local-law/arch_stnds_2.htm

▷ **Department of Defense *Archaeological Inventory Standards and Cost-Estimation Guidelines***
https://www.denix.osd.mil/denix/Public/ES-Programs/Conservation/Legacy/AISS/usacerl1.html

▷ **U.S. Fish and Wildlife Service Historic Preservation Home**
http://www.fws.gov/historicpreservation/crp/index.html

▷ **National Park Service Cultural Resource Management Guideline, *NPS-28***
http://www.cr.nps.gov/history/online_books/nps28/28contents.htm

▷ **Bureau of Land Management Heritage Resources Home**
http://www.blm.gov/heritage/

▷ **Federal Highway Administration State Cultural Resources Practices Database**
http://www.environment.fhwa.dot.gov/strmlng/es3stateprac.asp

▷ **National Association of Tribal Historic Preservation Officers**
http://www.nathpo.org/about.html

▷ **National Conference of State Historic Preservation Officers**
http://www.ncshpo.org/

▷ **National Association of State Archaeologists**
http://www.uiowa.edu/~osa/nasa/

▷ **Advisory Council on Historic Preservation**
http://www.achp.gov/

▷ **National Conference of State Legislatures Historic Preservation Legislation Database**
http://www.ncsl.org/programs/arts/statehist.htm

▷ **National Cooperative Highway Research Program Circular 347**
Managing Archaeological Investigations: A Synthesis of Highway Practice
http://onlinepubs.trb.org/onlinepubs/nchrp/nchrp_syn_347.pdf

For example, over the years working with tribal representatives in the U.S. and Micronesians in the Pacific Islands, we have become disenchanted with a strictly archaeological perspective on site composition, which often results in a site record that concentrates solely on physical expressions discernible by and readily understood by an archaeologist (soil color, artifacts, etc.). Our experience with the indigenous perspective has convinced us that a site is best recorded as part of its local environment, with reference to the water sources, trees, rocks, slopes, plants, and other natural and cultural features that may have had significance to its inhabitants.

On one recent six-week project, the senior author led a field survey team to a site previously recorded as a "historical trash scatter" which proved to be the remains of a small cabin. The lumber had decayed away, but the cabin was marked in outline by a rectangular scatter of cut nails. Two patches of broken pane glass, a doorknob, stovepipe flashing, and a hasp marked the locations of doors, windows, and a woodstove. The area around the cabin had a sparse scatter of fragmentary utilitarian glass, metal, and ceramic items with no clear ethnic signature. Typically, such a site is classified as "historic," which most professionals consider to be categorically different from a "Native American" site. Function was also not immediately evident. Three "treadle-style" honing wheel fragments were found, suggesting that steel blades were sharpened, but no other function was indicated.

Throughout this survey we were accompanied each day by two tribal traditionalists, one each from two nearby reservations, who supplied us with perspectives on Native American land use and other traditional activities. Their special knowledge led us to search for kinds of cultural resources that were new to our crew. For example, as a team we hunted wetlands and meadows for the ephemeral signs of traditional plant cultivation and land use.

All this was fascinating and important, but what really stood out in the whole experience were events related to the cabin. Because these knowledgeable people were with us as we visited and recorded the cabin site, they were able to recall stories that linked it to an Indian family who used the adjacent meadow for sheep grazing and who had actually filed a patent on the land in 1880 and used the cabin for more than 30 years. Their memories soon poured out in a cascade of useful interpretation. The treadle honing wheel was probably used to sharpen shears or knives for shearing or polling sheep; the parsimonious nature of the artifact assemblage and extremely fragmentary remains probably reflected the family's well-known economical lifestyle, and this probably also explained the absence of the woodstove (hauled away). Further, as we began to record we recognized that the site could not be understood without reference to the meadow, which on closer inspection we realized had been much bigger, diminished in the last 50 years by conifer invasion. None of this could be recognized without consultation with the descendant community, and in fact, an on-the-ground partnership during the field survey was absolutely critical.

We shouldn't limit our discussions with descendant communities to just those things that help us find and understand sites. These communities also have a real, palpable stake in the project outcome. In many cases an archaeologist taking on a new project will find him or herself inserted into a context with a lot of history, where the project has struck like a thunderclap and people on all sides respond in ways that are impossible to understand without a commitment to dedicated inquiry, listening, and understanding.

— See you out there!

References Cited

Aldenderfer, M. S., and C. A. Hale-Pierce
    1984   The Small-Scale Archaeological Survey Revisited. *American Archaeology* 4(1):4–5.

Aldenderfer, M. S., and H. D. G. Maschner, eds.
    1996   *Anthropology, Space, and Geographic Information Systems.* Oxford: Oxford University Press.

Aldenderfer, M. S., and F. J. Schieppati
    1984   To Be or Not To Be: The Small-Scale Archaeological Survey and Archaeological Research. *American Archaeology* 4(1):49–53.

Alexandria Archaeology
    1992   *Historic Preservation Plan for the City of Alexandria.* Prepared by Alexandria Archaeology, City of Alexandria, Virginia.

Allen, K. M. S., S. W. Green, and E. B. W. Zubrow, eds.
    1990   *Interpreting Space: GIS and Archaeology.* London: Taylor and Francis.

Ambler, J. R.
    1984   The Use and Abuse of Predictive Modeling in Cultural Resource Management. *American Archaeology* 4(2):140–146.

Ammerman, A. J.
    1981   Surveys and Archaeological Research. *Annual Review of Anthropology* 10:63–88.

Anderson, P. F.
    1995   GIS Modeling of Archaeological Sites in the Raccoon River Greenbelt, Dallas County, Iowa. Prepared for the Dallas County Conservation Board, Perry, Iowa. Electronic document, http://www.public.iastate.edu/~fridolph/archmod.html, accessed June 28, 2006.

Andresen, J., T. Madsen and I. Scollar, eds.
    1992   *Computing the Past: Computer Applications and Quantitative Methods in Archaeology.* Aarhus: Aarhus University Press.

Artz, J. A.
    1996   Cultural Response or Geological Process? A Comment on Sheehan. *Plains Anthropologist* 41(158):383–393.

Barcelo, J. A., I. Briz, and A. Vila, eds.
    1999   *New Techniques for Old Times.* BAR International Series 757. Oxford: Archaeopress.

Benchley, E.
    1976   An Overview of the Prehistoric Resources of the Metropolitan St. Louis Area. Washington, D.C.: Interagency Archaeological Services, Office of Archaeology and Historic Preservation, National Park Service.

Burtchard, G. C.
    2004   Environment, Prehistory & Archaeology of Mount Rainier National Park, Washington. Electronic document, http://www.nps.gov/archive/mora/ncrd/archaeology/contents.htm, accessed June 28, 2006.

Chang, K. C., ed.
    1968   *Settlement Archaeology.* Palo Alto: National Press.

Childress, J., and J. L. Chartkoff
    1966   *An Archaeological Survey of the English Ridge Reservoir in Lake and Mendocino Counties, California.* Robert E. Schenk Archives of California Archaeology, Number 23. San Francisco: Society for California Archaeology.

Collins, J. M., and B. L. Molyneaux
    2003   *Archaeological Survey.* Walnut Creek, CA: Altamira Press.

Dincauze, D. F., and J. W. Meyer
    1977   *Prehistoric Resources of East-central New England: A Preliminary Predictive Study.* Washington, D.C.: National Park Service, U.S. Dept. of the Interior.

Fredrickson, D. A.
    1949   *Appraisal of the Archeological Resources of New Melones Reservoir, Calaveras and Tuolumne Counties, California.* Smithsonian River Basin Survey Report. Washington, D.C.: Pacific Coast Division, River Basin Surveys, Smithsonian Institution.

Gallant, T. W.
   1986   "Background Noise" and Site Definition: A Contribution to Survey Methodology. *Journal of Field Archaeology* 13(4): 403–418.

Gagliano, S. W.
   1977   Cultural Resources Evaluation of the Northern Gulf of Mexico Continental Shelf, Vol. 1. Washington, D.C.: Interagency Archeological Services, National Park Service.

Garbarino, M. S.
   1977   *Sociocultural Theory in Anthropology*. New York: Holt, Rinehart and Winston.

Goldberg, P. and R. I. MacPhail
   2006   *Practical and Theoretical Geoarchaeology*. Malden, MA: Blackwell Publishing.

Harris, M.
   1968   *The Rise of Anthropological Theory*. New York: T. Y. Crowell.

Heizer, R. F.
   1974   Studying the Windmiller Culture. In *Archaeological Researches in Retrospect*, edited by G.R. Willey, pp. :179–204. Washington, D.C.: University Press of America.

Heizer, R. F. and J. Graham
   1967   *A Guide to Field Methods in Archaeology*. Palo Alto, CA: National Press.

Hester, T. R., H. J. Shafer, and K. L. Feder
   1997   *Field Methods in Archaeology*. Mountain View, CA: Mayfield Publishing.

Holliday, V. T.
   1997   *Paleoindian Geoarchaeology of the Southern High Plains*. Austin: University of Texas Press.

Judge, W. J., and L. Sebastian, eds.
   1988   *Quantifying the Present and Predicting the Past: Theory, Method, and Application of Archaeological Predictive Modeling*. Denver: U.S. Department of the Interior, Bureau of Land Management.

King, T. F.
   1966   *An Archaeological Survey of the Dos Rios Reservoir Region, Mendocino County, California*. Tucson: U.S. Department of the Interior, Western Archaeological Center, National Park Service.
   1975   *Fifty Years of Archaeology in the California Desert: An Archaeological Overview of Joshua Tree National Monument*. Tucson: U.S. Department of the Interior, Western Archaeological Center, National Park Service.

   1978   *The Archeological Survey: Methods and Uses*. Washington, D.C.: U.S. Department of the Interior, Heritage Conservation and Recreation Service.
   2003   Places That Count, Traditional Cultural Properties in Cultural Resource Management. Walnut Creek, CA: Altamira Press.

King, T. F. and P. P. Hickman
   1973   *The Southern Santa Clara Valley: A General Plan for Archaeology*. San Francisco: A. E. Treganza Anthropology Museum, San Francisco State University.

Kvamme, K.L.
   1992   Geographic Information Systems and Archaeology. In *Computer Applications and Quantitative Methods in Archaeology 1991*, G. Lock and J. Moffett, eds., pp. 77–84. BAR International Series S577. Oxford: British Archaeological Reports.

Larrabee, E. McM., and S. Kardas
   1966   *Archaeological Survey of Grand Coulee Dam National Recreation Area*. Part 1: *Lincoln County above Normal Pool*. Report of Investigations No. 38. Pullman, WA: Laboratory of Anthropology, Washington State University.

Lovis, W. A.
   1976   Quarter Sections and Forests: An Example of Probability Sampling in the Northeastern Woodlands. *American Antiquity* 41(3):364–371.

Lyon, E. A., II
   1982   New Deal Archaeology in the Southeast: WPA, TVA, NPS, 1934–1942. Unpublished Ph.D. dissertation. Baton Rouge, LA: Louisiana State University Agricultural and Mechanical College.

Matson, R. G. and W. D. Lipe
   1975   Regional Sampling: A Case Study of Cedar Mesa, Utah. In *Sampling in Archaeology*, edited by J. W. Mueller, pp. 124–143. Tucson: University of Arizona Press.

Mayne, A. J. C. and T. Murray
   2004   *The Archaeology of Urban Landscapes: Explorations in Slumland*. New York: Cambridge University Press.

Mehrer, M. W and K. L Wescott
   2006   *GIS and Archaeological Site Location Modeling*. Baton Rouge, FL: CRC Press.

Miller, J. J.
   2000   The Government Sector: Reforming the Archaeology Curriculum to Respond to New Contexts of Employment. In *Teaching Archaeology in the Twenty-First Century*, edited by S. J. Bender and G. S. Smith, pp. 69–72. Washington, D.C.: Society for American Archaeology.

Moratto, M. J.
  1972   A Study of Prehistory in the Southern Sierra Nevada Foothills, California. Ph.D. dissertation, Graduate School, Department of Anthropology, University of Oregon, Eugene.
  1976   New Melones Archaeological Projects—Stanislaus River, Calaveras and Tuolumne Counties, California, Phase VI. Report to Interagency Archaeological Services, San Francisco. (5 vols.)

Mueller, J. W.
  1974   *The Use of Sampling in Archaeological Survey.* Memoir 28, Washington, D.C.: Society for American Archaeology.

Mueller, J. W., ed.
  1975   *Sampling in Archaeology.* Tucson: University of Arizona Press.

Orser, C. E.
  2004   *Historical Archaeology.* Upper Saddle River, NJ: Prentice Hall.

Parker, P. I. and T. F. King
  1990   Guidelines for Evaluating and Documenting Traditional Cultural Properties. *National Register Bulletin 38.* Washington, D.C.: U.S. Department of the Interior, National Park Service, National Register, History and Education.

Pollard, A. M., ed.
  1999   *Geoarchaeology: Exploration, Environments, Resources.* Geological Society Special Publication no. 165. London: The Geological Society.

Rapp, G., Jr., and C. L. Hill
  1998   *Geoarchaeology: The Earth-Science Approach to Archaeological Interpretation.* New Haven: Yale University Press.

Redman, C. L.
  1982   Archaeological Survey and the Study of Mesopotamian Urban Systems. *Journal of Field Archaeology* 9(3):375–382.

Ruppe, R. J.
  1966   The Archaeological Survey: A Defense. *American Antiquity* 31:313–333.

Schiffer, M. B. and J. H. House, eds.
  1975   *The Cache River Archaeological Archaeological Project: An Experiment in Contract Archaeology.* Fayetteville, AR: Arkansas Archaeological Survey Research Series 8.
  1977   Cultural Resource Management and Archaeological Research: The Cache Project. *Current Archaeology* 18(1):43–53.

Schiffer, M. B., A. P. Sullivan, and T. C. Klinger
  1978   The Design of Archaeological Surveys. *World Archaeology* 10:1–28.

Sheehan, M. S.
  1995   Cultural Responses to the Altithermal or Inadequate Sampling? *Plains Anthropologist* 40(153):261–270.
  1996   "Cultural Responses to the Altithermal or Inadequate Sampling?" Reconsidered. *Plains Anthropologist* 41(158):395–397.

Smith, L. D.
  1977   *Archaeological Sampling Procedures for Large Land Areas: A Statistically Based Approach.* Albuquerque, NM: U.S.D.A. Forest Service.

Society for American Archaeology
  2003   SAA Membership Needs Assessment Survey. On-Line at http://ecommerce.saa.org/saa//staticcontent/staticpages/survey/index.cfm. Accessed 3 November 2006.

Squier, E. G., and E. H. Davis
  1848   *Ancient Monuments of the Mississippi Valley.* Smithsonian Contributions to Knowledge. Washington, D.C.: Smithsonian Institution.

Stephens, J. L.
  1841   *Incidents of Travel in Yucatan.* New York: Harper and Brothers.

Steward, J. H.
  1955   *Theory of Culture Change.* Urbana, IL: University of Illinois Press.

Talmadge, V., and O. Chesler
  1977   *The Importance of Small, Surface, and Disturbed Sites as Sources of Significant Archaeological Data.* Washington, D.C.: Interagency Archaeological Services, Office of Archaeology and Historic Preservation, National Park Service.

Thomas, D. H.
  1975   Nonsite Sampling in Archaeology: Up the Creek without a Site? In *Sampling in Archaeology,* J. W. Mueller, ed., pp. 61–81. Tucson: University of Arizona Press.

Turpin, S. A., ed.
  1991   *Papers on Lower Pecos Prehistory.* Studies in Archeology 8. Austin: Texas Archeological Research Laboratory.

Waters, M. R.
  2004   *Principles of Geoarchaeology: A North American Perspective,* 5th edition. Tucson: University of Arizona Press.

Weide, M. S.
    1974   *Archaeological Element of the California Desert Study.*
            Riverside, CA: Bureau of Land Management.

Wendorf, F.
    1962   *A Guide for Salvage Archaeology.* Santa Fe: Museum of New
            Mexico Press.

Wescott, K. L., and Brandon, R. J., eds.
    2000   *Practical Applications of GIS for Archaeologists.* Philadelphia:
            Taylor and Francis.

Wheeler, T.
    2000   Society for California Archaeology Membership Survey Report
            I. *Society for California Archaeology Newsletter* 34(4):20–
            22.
    2001   Society for California Archaeology Membership Survey Report
            II. *Society for California Archaeology Newsletter* 35(1):18–
            19.

White, L.
    1959   *The Evolution of Culture.* New York: McGraw-Hill.

Willey, G. R. and J. A. Sabloff
    1974   *A History of American Archaeology.* San Francisco: W. H.
            Freeman and Co.

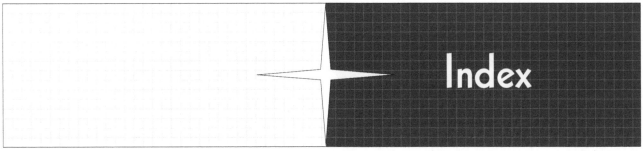

Note: page numbers in italics refer to illustrations.

Appendices

Appendix

Appendix A:

State Archaeological Survey Standards, Guidelines, and Forms Websites

Alabama
Alabama Historical Commission Policy for
Archaeological Survey and Testing in Alabama
http://www.preserveala.org/DOCUMENTS/PDF/revisedsurveyguidelines.pdf

Alaska
Standards and Guidelines for Investigating and
Reporting Archaeological and Historic Properties in Alaska
http://www.dnr.state.ak.us/parks/oha/hpseries/hp11.pdf

American Samoa
American Samoa Historic Preservation Office:
Survey and Inventory Program
http://ashpo.org/program.htm#survey

Arizona
The Arizona State Museum Procedures Manual for Arizona Antiquities Act Permits,
Records Management/Repository Requirements, and Archaeological Records Access
http://www.statemuseum.arizona.edu/profsvcs/permits/permit_manual.pdf

California
California Office of Historic Preservation:
Publications and Forms
http://ohp.parks.ca.gov/default.asp?page_id=1069

Colorado
Office of Archaeology and Historic Preservation, Colorado Historical Society:
Colorado Rules and Procedures
http://www.coloradohistory-oahp.org/publications/pubs/1308b.pdf

Connecticut
Procedures and Standards for the Issuance of Permits
for Archaeological Investigation on State Lands
http://www.chc.state.ct.us/ArchaeolPermits.htm

Delaware
Delaware State Historic Preservation Office Survey Forms:
Instructions and Data Coordination Guidance
http://history.delaware.gov/pdfs/DESHPO_SurveyFormsDataCoordGuidance.pdf

District of Columbia
DC Office of Historic Preservation:
Historic Preservation Laws and Regulations
http://www.planning.dc.gov/planning/cwp/view,A,1284,Q,637255.asp

Florida
Florida Division of Historical Resources:
Standards for Archaeological Reports
http://www.flheritage.com/preservation/compliance/laws/1a_46.pdf

Georgia
Georgia Council of Professional Archaeologists:
Standards and Guidelines for Archaeological Surveys
http://www.georgia-archaeology.org/GCPA/Georgia_standards.pdf

Appendix A: State Archaeological Survey Standards, Guidelines, and Forms Websites

▷ **Guam**
  Guam Office of Historic Preservation
  (note: server down at last attempt)
  http://ns.gov.gu/dpr/hrdhome.html

▷ **Hawaii**
  Review Process Resource Page:
  Standards and Guidelines for Professional Archaeologists
  http://www.hawaii.gov/dlnr/hpd/hpgreeting.htm

▷ **Idaho**
  Idaho State Historical Society and State Historic Preservation Office:
  Guidelines and Project Review Protocols
  http://www.idahohistory.net/shpo.html#anchor273075

▷ **Illinois**
  Illinois Historic Preservation Office:
  Guidelines for Archaeological Reconnaissance Surveys/Reports
  http://virtual.parkland.edu/ias/professional_resources/guidelines.html

▷ **Indiana**
  Division of Historic Preservation/Archaeology:
  Archaeology Mandates and Laws
  http://www.in.gov/dnr/historic/archeo_mandates.html

▷ **Iowa**
  State Historical Society of Iowa:
  Historic Preservation Subsection
  http://www.iowahistory.org/preservation/index.html

▷ **Kansas**
  Kansas State Historical Society:
  Archaeology Resources
  http://www.kshs.org/archeologists/index.htm

▷ **Kentucky**
  Kentucky Heritage Council (Office of Historic Preservation):
  Kentucky Historic Preservation Plan
  http://www.state.ky.us/agencies/khc/preservation_plan.htm

▷ **Louisiana**
  Division of Historic Preservation:
  Title 25 State Survey Guidelines and Regulations
  http://www.crt.state.la.us/archaeology/HOMEPAGE/25_Part_1.shtml

▷ **Maine**
  Historic Preservation Commission:
  Standards for Archaeological Work in Maine
  www.maine.gov/sos/cec/rules/90/94/089/089c812.doc

▷ **Maryland**
  Maryland Historical Trust (Office of Historic Preservation):
  Standards and Guidelines for Archaeological Investigations in Maryland
  http://www.marylandhistoricaltrust.net/arch-s&g.pdf

Massachusetts
Massachusetts Historical Commission:
Publications and Survey Forms
http://www.sec.state.ma.us/mhc/mhchpp/ppdhpp.htm

Michigan
State Historic Preservation Office:
Links and Contact Information
http://www.michigan.gov/hal/0,1607,7-160-17449_18638---,00.html

Minnesota
Minnesota Statutes, Chapter 138:
Archaeological Professional Licensing and Survey Standards
http://www.revisor.leg.state.mn.us/stats/138/

Mississippi
Mississippi Department of Archives and History, Mississippi State Historic Preservation
Office: Guidelines for Archaeological Investigations and Reports in Mississippi
http://www.mdah.state.ms.us/hpres/archguidelines.html

Missouri
State Historic Preservation Office:
Guidelines for Phase I Archaeological Surveys and Reports and Site Forms
http://www.dnr.mo.gov/shpo/Archaeology.htm

Montana
Montana Historical Society:
Guidelines and Procedures for Cultural Resource Review and Consultation
http://www.his.state.mt.us/shpo/archaeology/consultingwith.asp

Nebraska
State Historic Preservation Office:
Guidelines and Consultants List
http://www.nebraskahistory.org/histpres/archsurv.htm

Nevada
Department of Cultural Affairs, State Historic Preservation Office: Documentation Standards
for Historical Resources of State and Local Significance
http://dmla.clan.lib.nv.us/docs/shpo/siguidelines/documstnds.htm

New Hampshire
Division of Historical Resources:
Guidelines and Consultants List
http://www.state.nh.us/nhdhr/

New Jersey
State of New Jersey:
Guidelines for Phase I Archaeological Investigation
http://www.state.nj.us/dep/hpo/1identify/arkeoguide1.htm

New Mexico
Historic Preservation Division:
Standards for Permits, Survey, and Approved Consultants
http://www.nmhistoricpreservation.org/documents.html?&_recordnum=8

Appendix A: State Archaeological Survey Standards, Guidelines, and Forms Websites

▷ **New York**
New York State Museum:
Cultural Resource Standards Handbook
http://www.nysm.nysed.gov/research/anthropology/crsp/arccrsp_nyachb.html

▷ **North Carolina**
State Historic Preservation Office:
Guidelines for Preparation of Archaeological Survey Reports
http://www.arch.dcr.state.nc.us/ncarch/resource/crmguide.htm

▷ **North Dakota**
State Historical Society:
Links and Contacts
http://www.state.nd.us/hist/

▷ **Ohio**
Ohio Historic Preservation Office:
Archaeological Guidelines
http://www.ohiohistory.org/resource/histpres/toolbox/apothopo.html

▷ **Oklahoma**
State Historic preservation Office:
Site Forms and Consultants List
http://www.okhistory.org/shpo/consultants.htm

▷ **Oregon**
State Historic Preservation Office:
Archaeological Survey and Reporting Standards
http://www.oregon.gov/OPRD/HCD/SHPO/docs/arch_standards.pdf

▷ **Pennsylvania**
Cultural Resource Management in Pennsylvania:
Guidelines for Archaeological Investigations
http://www.phmc.state.pa.us/bhp/Inventories/ArchaeologyGuidelines.pdf

▷ **Rhode Island**
Historical Preservation and Heritage Commission:
Procedures for Registration and Protection of Historic Properties
http://www.preservation.ri.gov/review/regulations.php

▷ **South Carolina**
South Carolina Institute of Archaeology and Anthropology:
South Carolina Standards and Guidelines for Archaeological Investigations
http://www.cas.sc.edu/sciaa/pdfdocs/scs&g.pdf

▷ **South Dakota**
South Dakota State Historic Preservation Office:
Guidelines for Cultural Resource Surveys and Survey Reports
http://www.sdhistory.org/HP/R&C%20Guidelines.pdf

▷ **Tennessee**
Division of Archaeology:
Division Rules and Consultants List
http://www.state.tn.us/environment/arch/

Texas
Council of Texas Archaeologists:
Survey Standards
http://www.counciloftexasarcheologists.org/index.php?option=content&task=view&id=16&Itemid=55

Utah
Division of State History/Utah State Historical Society,
Archaeology Division: Standards and Guidelines
http://history.utah.gov/archaeology/laws_-_permits/internalprocedures.html

Vermont
State Historic Preservation Office:
Guidelines for Conducting Archaeology in Vermont
http://efotg.nrcs.usda.gov/references/public/VT/guidelines_for_conducting_arch.pdf

Virginia
Department of Historical Resources:
Guidelines for Conducting Cultural Resource Survey in Virginia
http://www.dhr.state.va.us/review/Survey_Manual_Web.pdf

Washington
Department of Archaeology and Historic Preservation:
Survey and Inventory Standards for Cultural Resource Reporting
http://www.oahp.wa.gov/documents/ExternalFinalFINAL_000.pdf

West Virginia
West Virginia Division of Culture and History: Guidelines for Phase I, II, and III
Archaeological Investigations and Technical Report Preparation
http://www.wvculture.org/shpo/techreportguide/guidelines.html

Wisconsin
Wisconsin Archeological Survey:
Guidelines for Public Archeology in Wisconsin
http://www.uwm.edu/Org/WAS/WASurvey/WASpubs.html

Wyoming
State Historic Preservation Office:
Standards and Guidelines for Class II and Class III Inventories
http://www.wy.blm.gov/cultural/forms/report-format.pdf

Appendix A: State Archaeological Survey Standards, Guidelines, and Forms Websites

Appendix B:
Example Archaeological Site Record Form

The site record form offered in this appendix was produced just for this volume and was put together after examining a number of forms used by various states and federal agencies in the U.S. It is not meant to replace those forms, but to provide a synthesis and distillation of typical examples.

Many state offices of historic preservation, tribal historic preservation officers, state archaeologists, and individual federal agencies have a big inventory of site record forms, subforms, and attachments which they prescribe for particular site types and circumstances. A few of these are posted on-line with manuals.

Our form—reproduced here—like all forms, is rigid, and therefore accommodating in some respects and contrary in others. Even if you are compelled to use a difficult form or one maladapted to your circumstance, be prepared to use it thoughtfully and creatively.

Archaeological Site Record

Sheet _____ of _____

Historic or Common Name(s): _____

State Code: _____

Site Field Designation#(s): _____

NRHP Code: _____

Site Trinomial(s): _____

Other Code(s): _____

USGS Quad(s):(15') _____

(7.5') _____

T/R, Section #(s): _____

Aliquot Part (list ¼ of ¼ of ¼, if possible): _____

UTM Location

UTM Zone/Datum: _____ Site Datum: _____ mE / _____ mN

UTM Corners: NW: _____ mE / _____ mN   NE: _____ mE / _____ mN

SW: _____ mE / _____ mN   SE: _____ mE / _____ mN

1. Site Location (legal description, driving directions, narrative location):

_____

_____

_____

_____

_____

_____

_____

_____

_____

_____

_____

_____

☐ *Continuation Sheet*

Project Name and Lead Agency: _____

Report Reference: _____

_____

_____

Archaeological Site Record

Sheet _____ of _____

Historic or Common Name(s): _____   State Code: _____

Site Field Designation#(s): _____   NRHP Code: _____

Site Trinomial(s): _____   Other Code(s): _____

NRHP Resource Type: Building ☐   Structure ☐   Object ☐   Site ☐   District ☐   Element of District ☐   Other ☐ (explain)

Component(s) and Age Estimate(s)   Single Component: | Yes | No | Multi-Component: | Yes | No |

Historic ☐ Time Markers Observed: _____

    Feature Illustrations ☐ _____
    Artifact Illustrations ☐ _____

Prehistoric☐ Time Markers Observed: _____

    Feature Illustrations ☐ _____
    Artifact Illustrations ☐ _____

Contact ☐ Time Markers Observed: _____

    Feature Illustrations ☐ _____
    Artifact Illustrations ☐ _____

Cultural Attribution (describe each component's attribution and data sets used to make determination):

_____

_____

_____

_____

☐ *Continuation Sheet*

Narrative Description of Site:

_____

_____

_____

_____

_____

_____

_____

_____

_____

_____

☐ *Continuation Sheet*

Archaeological Site Record

Sheet _____ of _____

Historic or Common Name(s): _____

State Code: _____

Site Field Designation#(s): _____

NRHP Code: _____

Site Trinomial(s): _____

Other Code(s): _____

Horizontal Dimensions (scale): Maximum Length: _____ ( ) along Bearing: _____

Method of Determination
Pacing ☐
GPS ☐
Tape ☐

Maximum Width: _____ ( ) along Bearing: _____

Site Area Estimate: _____ ^{2} ( ) along Bearing _____

Vertical Dimensions (scale): Maximum Depth: _____ ( ) **Stratified Deposits?:** | Yes | | No |

Method of Determination: Auger ☐ Shovel Test ☐ Cutbank ☐ Other ☐ (explain)

_____

_____

_____

2. Cultural Constituents (describe and quantify all observed cultural materials):

_____

_____

_____

_____

_____

_____

_____

_____

☐ *Continuation Sheet*

3. Cultural Features (inventory, describe associations, indicate size, and interpret all observed features):

_____

_____

_____

_____

☐ *Continuation Sheet*

4. Sampling Limitations (restricted access, pavement/buildings, natural obstructions, disturbance; explain):

_____

_____

_____

_____

☐ *Continuation Sheet*

Archaeological Site Record

Sheet _____ of _____

Historic or Common Name(s): _____

State Code: _____

Site Field Designation#(s): _____

NRHP Code: _____

Site Trinomial(s): _____

Other Code(s): _____

Elevation: _____ ( ) Slope: _____ ( ) Aspect: _____

---

5. Description of Local Biological Environment (habitat zones, associated plant and animal species, nearest water and water type):

_____

_____

_____

_____

_____

_____

_____

_____

☐ *Continuation Sheet*

---

6. Description of Local Geophysical Environment (landscape setting, landform, geology, erosion/deposition, sediments/soils):

_____

_____

_____

_____

_____

_____

_____

_____

☐ *Continuation Sheet*

---

7. Site Condition (note positive factors such as intact deposits and exant features, and problems including looting, erosion, fire, bulldozing, etc.):

_____

_____

_____

_____

_____

_____

☐ *Continuation Sheet*

Archaeological Site Record

Sheet _____ of _____

Historic or Common Name(s): _____

State Code: _____

Site Field Designation#(s): _____

NRHP Code: _____

Site Trinomial(s): _____

Other Code(s): _____

---

6. Interpretation (make a case and specify supporting evidence for site type, site function, site significance):

_____

_____

_____

_____

_____

_____

_____

_____

_____

_____

_____

_____

_____

_____

_____

_____

_____

_____

_____

_____

☐ *Continuation Sheet*

STP Log Attached: Yes | No   Feature Map Attached: Yes | No

Artifact Illustration(s) Attached: Yes | No   Photo Record Attached: Yes | No

Continuation Sheet(s) Attached: Yes | No   Artifact Log Attached: Yes | No

Sketch Map Attached: Yes | No   Location Map Attached: Yes | No

---

Recorded By (name/address/affiliation): _____

_____

Owner/Address/Other Contact Information: _____

_____

Appendix C:

*Example Archaeological
Survey Tracking and
Organization Forms*

Daily Survey Team Report

Sheet _____ of _____

Project: _____   Team: _____

Site #(s): _____   _____

GPS Datalog File #: _____   Date: _____

Coverage

Coverage Map Attached: | Yes | No |

USGS Quad(s): _____

_____

T/R, Section #(s): _____

UTM Corners: NW: _____   NE: _____

SW: _____   SE: _____

Survey Area Description: _____

_____

_____

_____

Site(s) Encountered: _____   Acres: _____

Isolate(s) Encountered: _____   Person-Days: _____

Site Log Attached: | Yes | No | Isolate Log Attached: | Yes | No | Acres/PPD: _____

Logistics

Conditions affecting coverage
(weather, surface visibility,
topography, vegetation): _____

_____

_____

Vehicle 1: _____   Mileage: _____   Crew Chief: _____

Vehicle 2: _____   Mileage: _____   Signed: _____

Archaeological Site Log Sheet

Sheet _____ of _____

Project: _____

Site #	UTM Coordinates	@ datum	Site Description	Team	Date
	E: N:				
	E: N:				
	E: N:				
	E: N:				
	E: N:				
	E: N:				
	E: N:				
	E: N:				
	E: N:				
	E: N:				
	E: N:				
	E: N:				
	E: N:				
	E: N:				
	E: N:				
	E: N:				
	E: N:				
	E: N:				

Site Documentation Log Sheet*

Sheet _____ of _____

Project: _____   Team: _____

Site #	Site Record	Sketch Map	Feature Map(s)	Site Location Map	Artifact Record	Artifact Illustration	Site Photographs	Photo Record	Permanent Datum	Packet Filed	Quality-Control Check

* initial and date on completion of each task and form.

Isolated Find Log Sheet

Sheet _____ of _____

Project: _____

Isolate #	UTM Coordinates	Description	Team	Date	Collected
	E: N:				Y / N
	E: N:				Y / N
	E: N:				Y / N
	E: N:				Y / N
	E: N:				Y / N
	E: N:				Y / N
	E: N:				Y / N
	E: N:				Y / N
	E: N:				Y / N
	E: N:				Y / N
	E: N:				Y / N
	E: N:				Y / N
	E: N:				Y / N
	E: N:				Y / N
	E: N:				Y / N
	E: N:				Y / N
	E: N:				Y / N
	E: N:				Y / N
	E: N:				Y / N

STP Inventory and Log Sheet

Sheet _____ of _____

Project: _____ Site: _____ STP Size: _____ Team: _____

Transect #: _____ Screen: Yes [ ] No [ ] _____

STP #s: _____ Screen Size: _____ Date: _____

STP #	UTM Coordinates	Constituents	Soils (Description/Munsell)
	E: N:		
	E: N:		
	E: N:		
	E: N:		
	E: N:		
	E: N:		
	E: N:		
	E: N:		
	E: N:		
	E: N:		
	E: N:		
	E: N:		
	E: N:		
	E: N:		
	E: N:		

Artifact Log Sheet

Sheet _____ of _____

Project: _____

Team: _____

Site # (s): _____

_____

Locus: _____

Date: _____

Artifact #	UTM Coordinates	Distance/Bearing (Datum)	Description	Collected
	E: N:			Y / N
	E: N:			Y / N
	E: N:			Y / N
	E: N:			Y / N
	E: N:			Y / N
	E: N:			Y / N
	E: N:			Y / N
	E: N:			Y / N
	E: N:			Y / N
	E: N:			Y / N
	E: N:			Y / N
	E: N:			Y / N
	E: N:			Y / N
	E: N:			Y / N
	E: N:			Y / N
	E: N:			Y / N
	E: N:			Y / N

Photographic Record

Sheet _____ of _____

Project: _____ Team: _____ Date(s): _____

Camera #: _____ Make/Model: _____ MP: _____ Lens: _____ Filter: _____

Image #	UTM Coordinates	Subject (Site/Feature/Artifact/Other)	Facing (Bearing)	Image Specs	Card
	E: N:				
	E: N:				
	E: N:				
	E: N:				
	E: N:				
	E: N:				
	E: N:				
	E: N:				
	E: N:				
	E: N:				
	E: N:				
	E: N:				
	E: N:				
	E: N:				
	E: N:				
	E: N:				
	E: N:				
	E: N:				

Scale and Arrow Templates

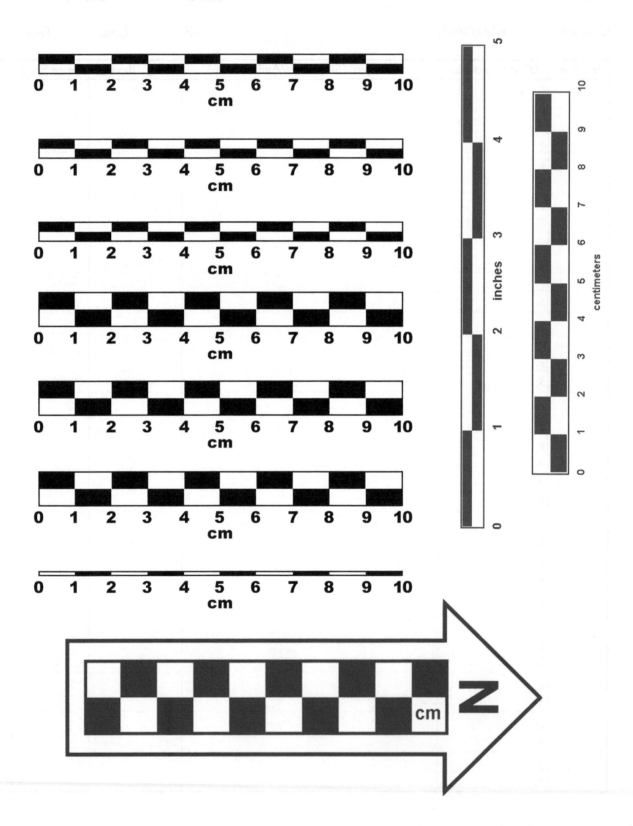

Artifact Collection Tags

Surface Collection	Surface Collection	Surface Collection	Surface Collection
Site:	Site:	Site:	Site:
Locus:	Locus:	Locus:	Locus:
Art#:	Art#:	Art#:	Art#:
Description:	Description:	Description:	Description:
Date:	Date:	Date:	Date:
Team:	Team:	Team:	Team:

Surface Collection	Surface Collection	Surface Collection	Surface Collection
Site:	Site:	Site:	Site:
Locus:	Locus:	Locus:	Locus:
Art#:	Art#:	Art#:	Art#:
Description:	Description:	Description:	Description:
Date:	Date:	Date:	Date:
Team:	Team:	Team:	Team:

Surface Collection	Surface Collection	Surface Collection	Surface Collection
Site:	Site:	Site:	Site:
Locus:	Locus:	Locus:	Locus:
Art#:	Art#:	Art#:	Art#:
Description:	Description:	Description:	Description:
Date:	Date:	Date:	Date:
Team:	Team:	Team:	Team:

Surface Collection	Surface Collection	Surface Collection	Surface Collection
Site:	Site:	Site:	Site:
Locus:	Locus:	Locus:	Locus:
Art#:	Art#:	Art#:	Art#:
Description:	Description:	Description:	Description:
Date:	Date:	Date:	Date:
Team:	Team:	Team:	Team:

Appendix D:

Example Archaeological Survey Field Notes

The field note format shown in this appendix is the one used routinely by the senior author to guide students and survey staff in organizing their notes. The note pages shown are from the senior author's own notebooks. Others use different formats, often designed specifically for a given project or area. There is no "right" format; the point simply is to record what you do, observe, and interpret in a clear, complete, and concise manner that is likely to be understandable by a reader picking up your notebook at some time in the distant future. Think about that reader as you organize and prepare your notes. He or she will thank you for it.

This frontspiece can be glued to the inside cover of quadrille-ruled field notebooks handed out to students and staff on each project.

Project Name:

Project Location:

If found, please return to:

Recommended Field Note Format

Your notes should serve as a daily journal of your scientific activities and observations:

Start on a new page each day. The header should contain:

a) Date
b) Location/Name of Investigation
c) Other conditions (weather, lighting, wind)

d) Your first notes-of-the-day should be done during the first meeting of the day, during which project leaders should describe the status of the investigation, weekly and daily goals, and your assignment. It should also be clear how your daily assignment fits into these goals. As you progress through the day, take some time to record:

e) Kind of activity (Planning, Survey, STP Excavation, Site Recording, etc.)
f) Crew Chief (use initials)
g) Co-workers (use initials)
h) Your Assignment

i) The bulk of your notes should pertain to ongoing activities. Make sure that these notes are enriched by well-labeled illustrations, including a sketch map of the general location of the survey area, specific transect location including UTM coordinates, a list of sites and isolates encountered, and illustrations of artifacts and features encountered.

j) Use evening time reserved for note-taking to clean up your notes, redraw illustrations, check with co-workers, review documentation, and follow-up with other observations and interpretations.

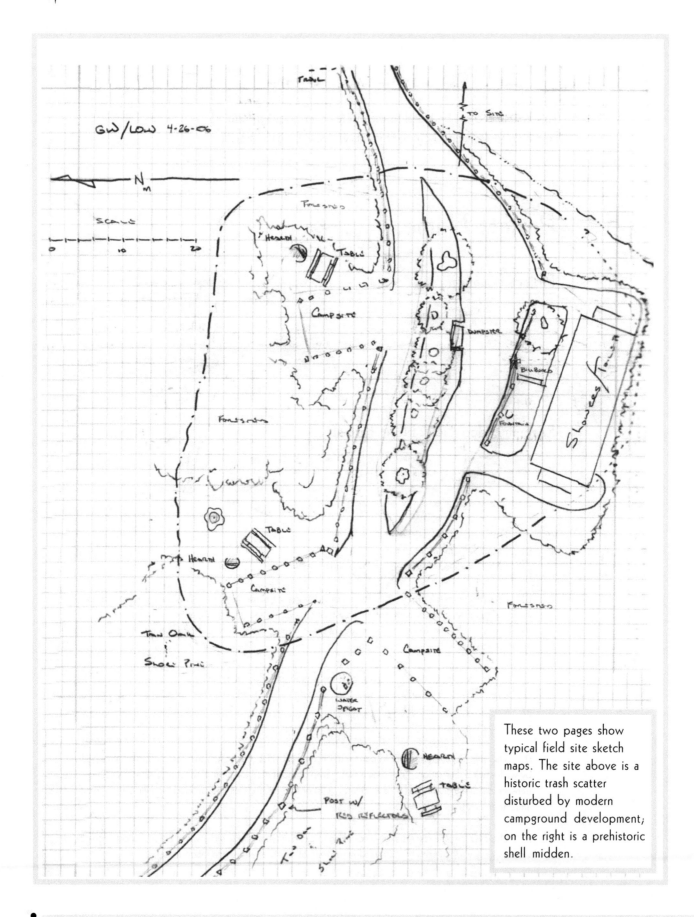

These two pages show typical field site sketch maps. The site above is a historic trash scatter disturbed by modern campground development; on the right is a prehistoric shell midden.

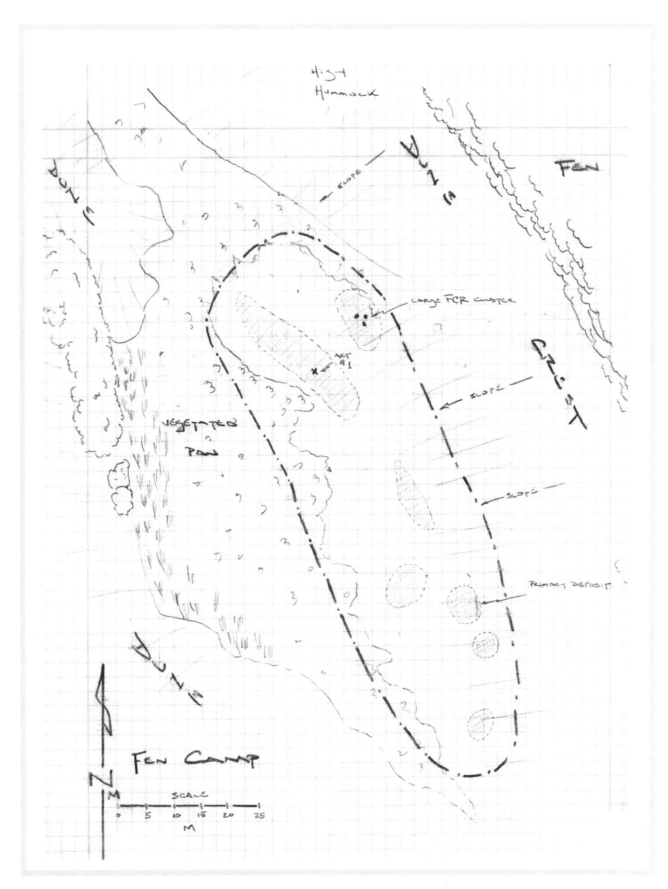

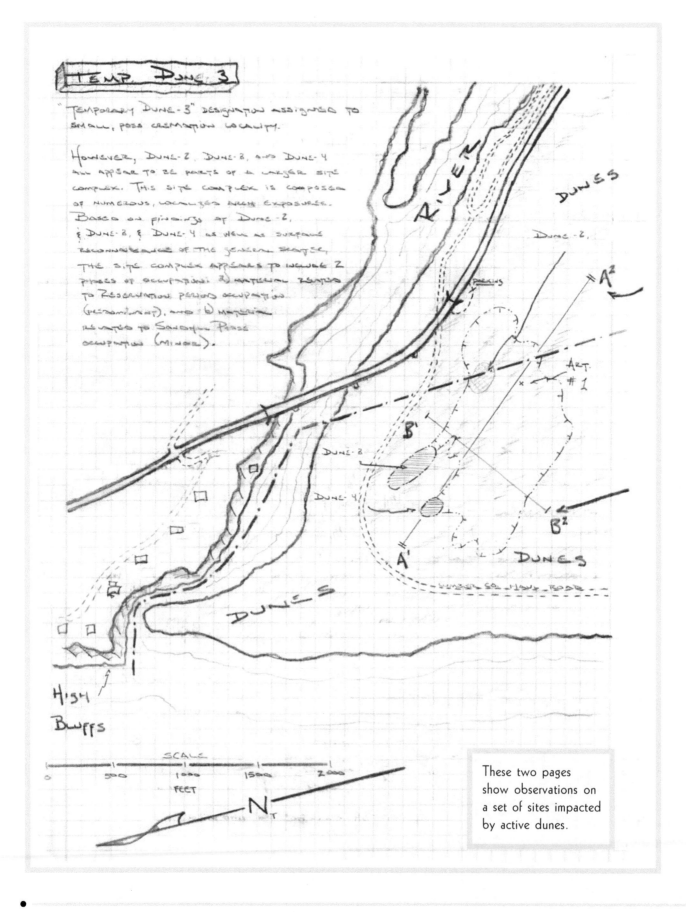

These two pages show observations on a set of sites impacted by active dunes.

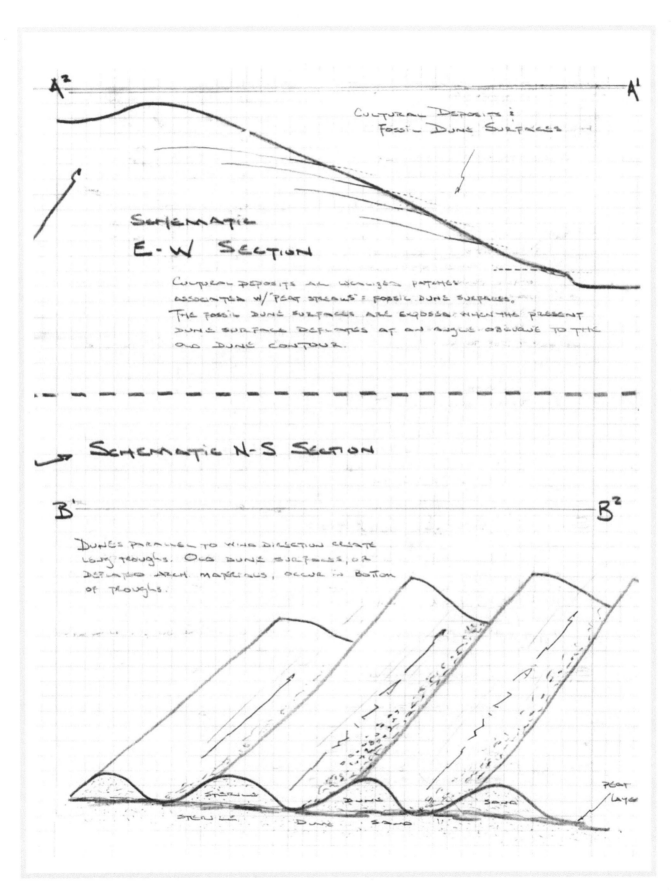

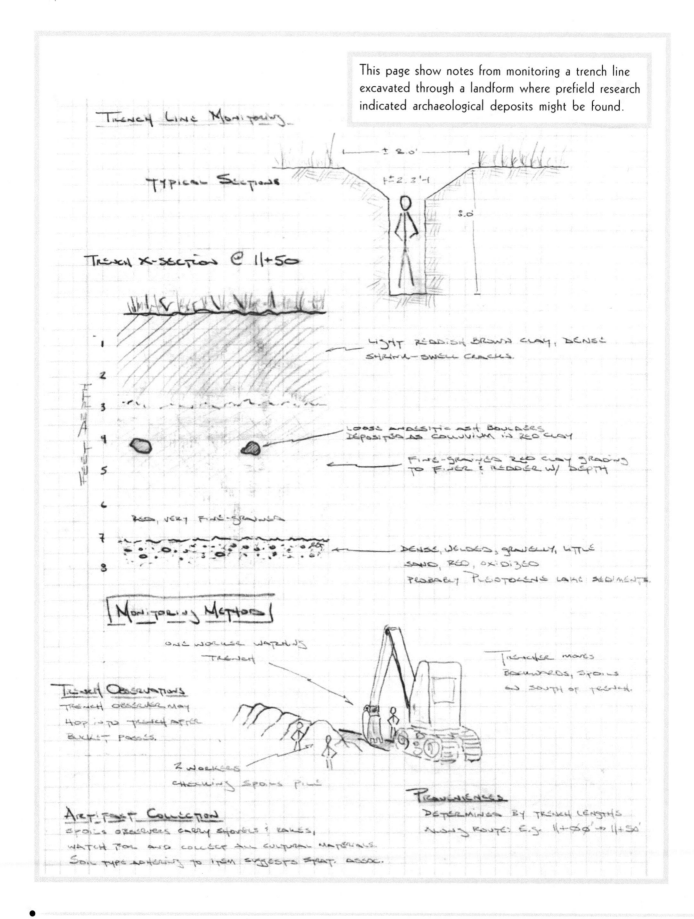

This page show notes from monitoring a trench line excavated through a landform where prefield research indicated archaeological deposits might be found.

TRENCH LINE MONITORING

TYPICAL SECTIONS

± 8.0'

± 2.3'

3.0'

TRENCH X-SECTION @ 11+50

1

2

3

LIGHT REDDISH BROWN CLAY, DENSE SHRINK-SWELL CRACKS.

4

LOOSE ANDESITIC ASH BOULDERS DEPOSITED AS COLLUVIUM IN RED CLAY

5

FINE-GRAINED RED CLAY GRADING TO FINER & REDDER W/ DEPTH

6

RED, VERY FINE-GRAINED

7

8

DENSE, WELDED, GRAVELLY, LITTLE SAND, RED, OXIDIZED PROBABLY PLEISTOCENE LAKE SEDIMENTS.

MONITORING METHODS

ONE WORKER WATCHING TRENCH

TRENCHER MOVES BACKWARDS, SPOILS GO SOUTH OF TRENCH.

TRENCH OBSERVATIONS
TRENCH OBSERVER MAY HOP INTO TRENCH AFTER BUCKET PASSES.

2 WORKERS CHECKING SPOILS PILE

ARTIFACT COLLECTION
SPOILS OBSERVERS CARRY SHOVELS & RAKES, WATCH FOR AND COLLECT ALL CULTURAL MATERIALS. SOIL TYPE ADHERING TO ITEM SUGGESTS STRAT. ASSOC.

PROVENIENCE
DETERMINED BY TRENCH LENGTHS ALONG ROUTE. E.g. 11+00' TO 11+50'

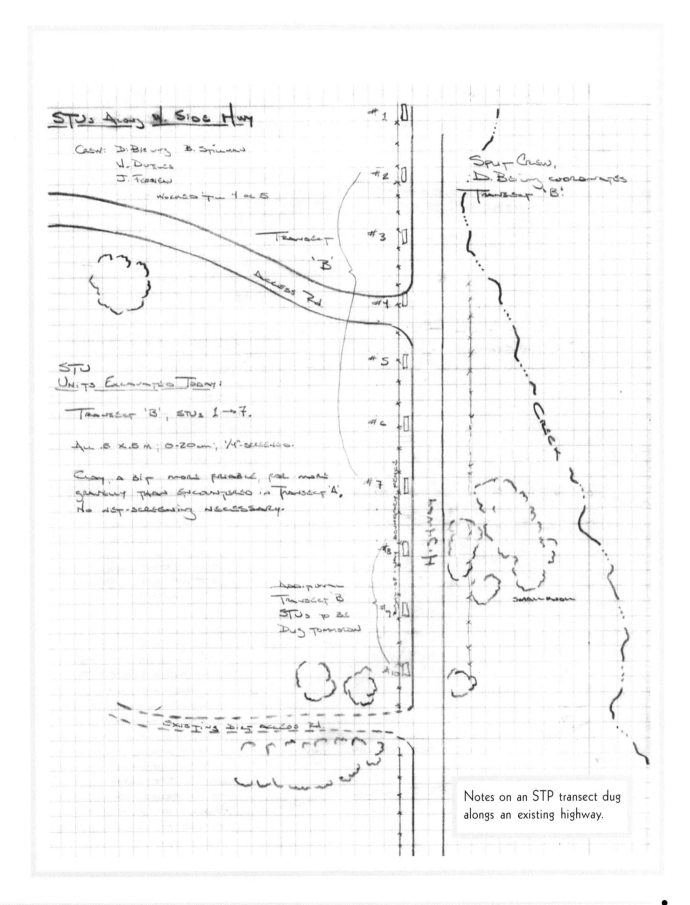

Notes on an STP transect dug alongs an existing highway.

These two pages show the notes taken while inspecting a deep stream cut in a valley designated for reservoir construction. Observations made this day led to subsequent geoarchaeological exploration which found extensive buried prehistoric archaeological deposits.

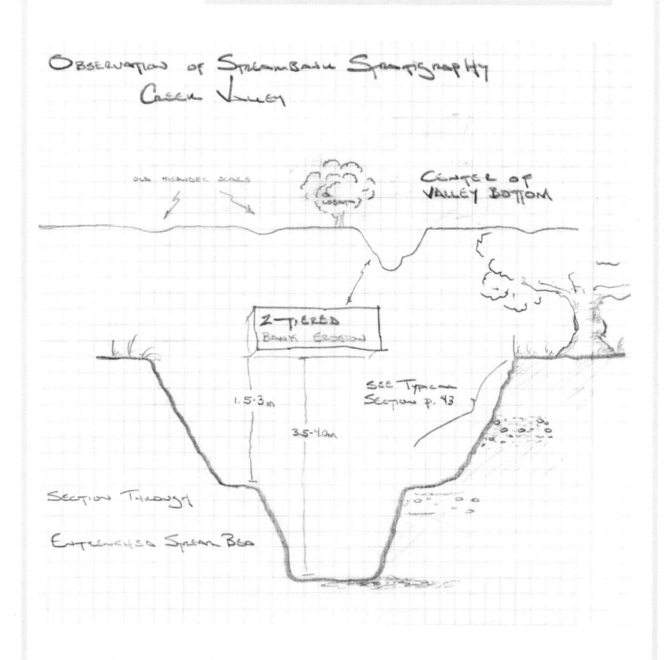

OBSERVATION OF STREAMBANK STRATIGRAPHY
CREEK VALLEY

OLD MEANDER SCARS

CENTER OF VALLEY BOTTOM

2-TIERED BANKS EROSION

1.5-3m

3.5-4.0m

SEE TYPICAL SECTION p. 43

SECTION THROUGH

ENTRENCHED STREAM BED

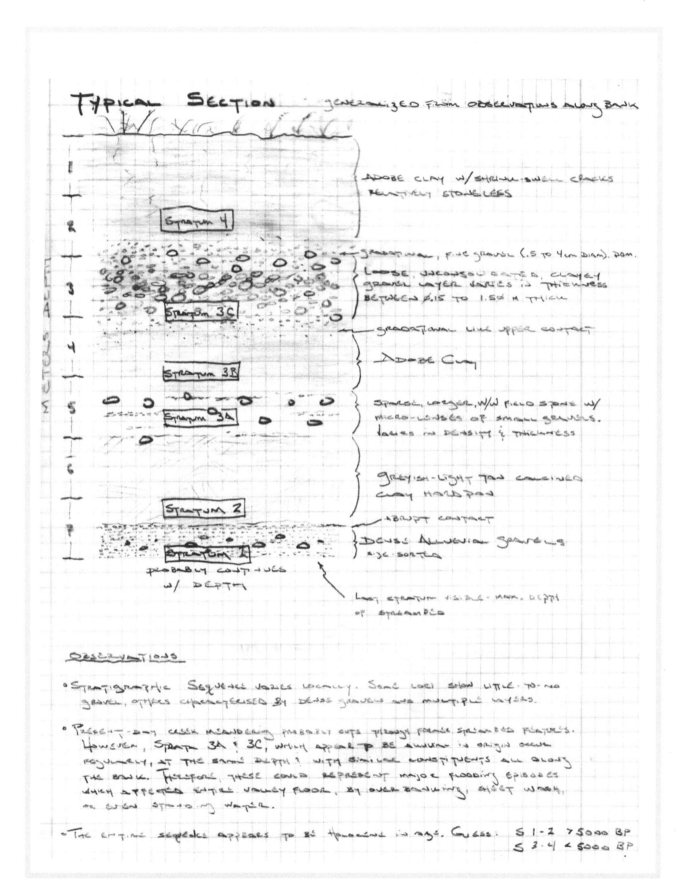

Field notes can be a forum for documentation of all your field studies, including apparent tangents. Some of these tangents can be counter-intuitive but surprisingly productive exercises, and some can be blind alleys. These two pages show the notes from a two-day study of the non-human contribution to bone distribution in a coastal dune field. This helped exclude certain features as cultural.

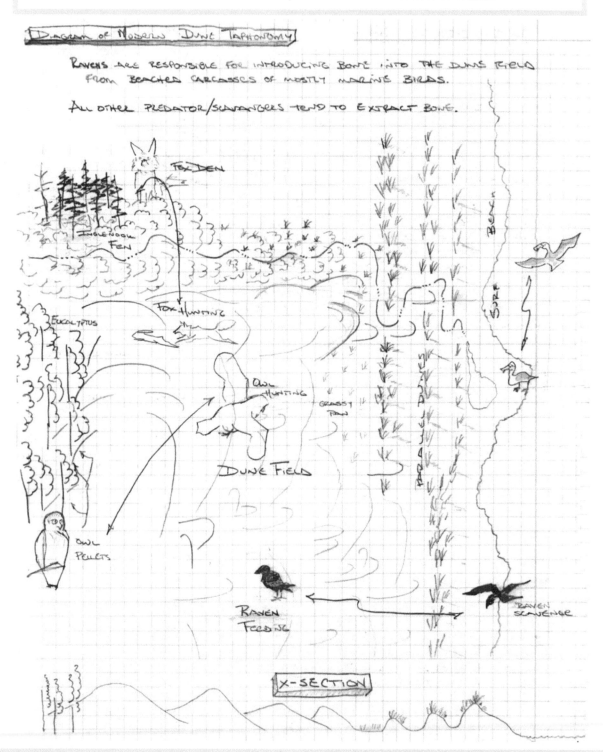

Dune Field Taphonomy

Modern dune field fauna constrained by narrow belts of poor vegetation.

However, there are lots of _Lepus C._ around bunch grass and vegetated pans. Occasional deer trekking through. Skunks, raccoon, poss. martin/weasel, small birds, deer, in vicinity of fen. Mid-size buteos and large accipiters prowling dune field appear to be hunting mice. Rabbits trek between grass patches and vegetated hummocks. Probably hunted by fox, coyote, owls, but as yet haven't observed them.

Beached marine fauna is rarely tossed or blown past first high (and vegetated) dune. These tend to collect around beached driftwood and are often "clumped" around a few objects on a given stretch of beach.

Remains of hunting and scavenging contained within the terrestrial (dune fields) ecosystem are scattered and unrepresentative. Roost trees favored by birds of prey are restricted to the eastern margin of the dune field (although accipiters were observed perched on tall, isolated vegetated hummocks). Therefore, regurgitated bone is probably clumped along the east side of the field. This may explain the scarcity of _Dipodomys_, _Peromyscus_, and _Perognathus_-sized bone in the dune field. However, _Thomomys_ mandibles and frontal/maxilla occur scattered around some vegetated patches. Mammalian predators probably carry-off their catch or pass the bone outside the dune field, because the current situation provides no suitable places for dens.

Most common constituent of modern taphonomy is bones of shore birds, some small fish bone, and a few pieces of crab shell scattered all across dune field. These bones are most common on dunes nearer the surf (w/i 500m), and are clustered high up the slopes of large parallel dunes on the long, low, west slope.

These bones appear to be feeding refuse of ravens (and possibly tidal turkeys). Apparently, the birds scavenge beached carcasses and fly these back to large, open dune faces to fight-over and devour the meat. Wing and limb elements are most common among the bird bone.

Ravens were observed to make for the surf from trees to the east once or twice today. They probably time the scavenging flights to the low tide. Will check tomorrow.

2:23 → Saw ravens patrolling coast ≈9:00 A.M.

About the Authors

Gregory G. White is director of the Archaeological Research Program at California State University, Chico, and a member of the board of the Society for California Archaeology.

Thomas F. King is an archaeological consultant and workshop leader for SWCA Environmental Consultants and author of half a dozen archaeology books including *Cultural Resource Law and Practice*, *Amelia Earhart's Shoes*, and *Doing Archaeology*.

T - #0703 - 101024 - C0 - 279/216/11 - PB - 9781598740097 - Gloss Lamination